Cambridge Elements ☰

Elements in the Philosophy of Mind
edited by
Keith Frankish
The University of Sheffield

THE COMPUTATIONAL THEORY OF MIND

Matteo Colombo
Tilburg University

Gualtiero Piccinini
University of Missouri – St. Louis

CAMBRIDGE
UNIVERSITY PRESS

Shaftesbury Road, Cambridge CB2 8EA, United Kingdom

One Liberty Plaza, 20th Floor, New York, NY 10006, USA

477 Williamstown Road, Port Melbourne, VIC 3207, Australia

314–321, 3rd Floor, Plot 3, Splendor Forum, Jasola District Centre, New Delhi – 110025, India

103 Penang Road, #05–06/07, Visioncrest Commercial, Singapore 238467

Cambridge University Press is part of Cambridge University Press & Assessment, a department of the University of Cambridge.

We share the University's mission to contribute to society through the pursuit of education, learning and research at the highest international levels of excellence.

www.cambridge.org
Information on this title: www.cambridge.org/9781009454070

DOI: 10.1017/9781009183734

First published 2023

A catalogue record for this publication is available from the British Library

ISBN 978-1-009-45407-0 Hardback
ISBN 978-1-009-18372-7 Paperback
ISSN 2633-9080 (online)
ISSN 2633-9072 (print)

The Computational Theory of Mind

Elements in the Philosophy of Mind

DOI: 10.1017/9781009183734
First published online: November 2023

Matteo Colombo
Tilburg University

Gualtiero Piccinini
University of Missouri – St. Louis

Author for correspondence: Matteo Colombo, m.colombo@uvt.nl

Abstract: The Computational Theory of Mind says that the mind is a computing system. It has a long history going back to the idea that thought is a kind of computation. Its modern incarnation relies on analogies with contemporary computing technology and the use of computational models. It comes in many versions, some more plausible than others. This Element supports the theory primarily by its contribution to solving the mind-body problem, its ability to explain mental phenomena, and the success of computational modelling and artificial intelligence. To be turned into an adequate theory, it needs to be made compatible with the tractability of cognition, the situatedness and dynamical aspects of the mind, the way the brain works, intentionality, and consciousness.

Keywords: mind, computation, representation, brain, cognition

ISBNs: 9781009454070 (HB), 9781009183727 (PB), 9781009183734 (OC)
ISSNs: 2633-9080 (online), 2633-9072 (print)

Contents

1 Introduction

The computational theory of mind (CTM) says that the mind is a computing system. Since the nervous system is the main biological organ of the mind in the animal kingdom, CTM implies that the nervous system, and the brain in particular, implements mental computations.[1] To unpack CTM and turn it into a fully fledged theory, several questions should be addressed: what are computing systems and how do they differ from systems that do not compute? Are all aspects of all minds computational? Do mental computations process representations? What kinds of representations? How do different computing processes produce and explain different kinds of mental phenomena? How do these processes change over time and adapt to different tasks and environments? What are the arguments for and against CTM?

Different philosophers and cognitive scientists have answered these questions in different ways, resulting in substantially different formulations of CTM. So, it would be more accurate to speak of computational theories of mind (in the plural), rather than *the* computational theory of mind. Perhaps even more accurately, CTM is a family of research programs consisting of attempts to model and explain mental capacities computationally (Miłkowski 2018).

Here, our objective is to clarify CTM, outline the diversity of computational approaches to the mind, and evaluate central arguments for and against CTM. We begin with a historical survey of some of the insights that have informed contemporary formulations of CTM (Section 2). After distinguishing different notions of *computation* (Section 3), we explain the risk of trivialising CTM and scout various accounts of how computation can be physically implemented (Section 4). Finally, we lay out the main arguments in favour of CTM (Section 5) as well as outstanding challenges and future vistas (Section 6).

2 Historical Background

Standard histories of CTM begin in the middle of the twentieth century, when behaviourism was the dominant approach in psychology in the USA. According to behaviourism, the proper target for scientific psychology was the observable behaviour of humans and other animals and the ways environmental stimuli affect it. The inner workings of the mind were out of bounds. Around the 1950s, a 'cognitive revolution' took place, which reinstated internal mental capacities, such as perception, problem solving, and language processing, as appropriate targets of psychological research. This renewed interest was fuelled by

---

[1] Some argue that plants, fungi, and single-celled organisms – which lack nervous systems in the standard sense – have at least some mental capacities (e.g., Lyon et al. 2021; Segundo Ortín & Calvo 2022), and therefore, given CTM, they implement computations (e.g., Kirkpatrick 2022).

developments in computer science; a major impetus for this new discipline, which was later called *cognitive science*, was the advent of the electronic digital computer.

Though standard, this account distorts and oversimplifies the origins of CTM. The view that mental capacities have something to do with computation predates the invention of digital computers; the rise of CTM is more nuanced than the standard account might suggest. To better understand CTM, we shall now take a brief historical tour, where we highlight the pedigree of various insights underlying CTM, particularly the ideas that some physical systems can implement computations and that mental capacities can be explained computationally.

2.1 Computation from Llull to Lovelace

In the Middle Ages, mathematical calculation methods were developed. Muḥammad ibn Mūsā al-Khwārizmī (died ca. 850) was a Persian polymath who developed several techniques for solving algebraic equations. The word 'algorithm', which means mathematical procedure guaranteed to solve every instance of a general problem, derives from his name. Specialised techniques for automatic, mechanical, and linguistic reasoning were also developed in the European Late Middle Ages and the Renaissance (Uckelman 2018). The logician and theologian Ramon Llull (1232/33–1315/16) wanted to devise a mechanical system for argumentation, which could demonstrate the indisputable correctness of Christian theology. Llull tried to implement his system by means of concentrically arranged paper discs containing words or letters from a finite alphabet, which could be rotated to return new combinations of words and letters. Although the workings of Llull's paper discs were simple and prone to error, Llull's wheels are an early attempt to capture an intelligent activity such as argumentation by means of a computation procedure and implement such a procedure in physical artefacts. In this example, reasoning is performed by manipulating an artefact, setting the stage for future attempts to formalise human reasoning in terms of simple step-by-step procedures and to implement such procedures in concrete, physical devices.

Llull's work influenced René Descartes (1596–1650). Inspired by technological advances like clockworks, fountains, and automata such as bell strikers in clock towers, Descartes developed detailed accounts of perception, action, memory, and emotion in terms of mechanical processes that respond to external stimulations by following a sequence of pre-determined operations. And yet, Descartes believed that thought and reasoning are distinctive human mental capacities that cannot be decomposed into pre-determined mechanical operations. Unlike other capacities, thought and reasoning produce genuinely novel behaviour that would elude a mechanistic account. The question of how mechanical, computational processes can account for novelty and creativity remains a concern in contemporary debates about CTM (Isaac 2018a).

Like Descartes, Thomas Hobbes (1588–1679) believed that many mental capacities can be explained mechanically. Unlike Descartes, Hobbes explicitly identified thought and reasoning with computation, which he understood as the arithmetical operations of addition and subtraction. This idea – that thought is a form of computing – influenced Gottfried Wilhelm von Leibniz (1646–1716). Similar to Llull, Leibniz designed a device for general reasoning, which would follow a system of formal operations for (re)combining simple linguistic symbols. As with Llull's paper discs, implementing Leibniz's device in concrete physical mechanisms proved challenging, as the concrete mechanisms would not always follow the rules they were meant to follow and often produced errors (Isaac 2018a, Section 2).

The mathematician George Boole (1815–64) studied what are now known as *Boolean algebras,* where the values of the variables are just the truth values *true* and *false,* often written as *1* and *0*. Operations on variables that can take two values later became the building blocks of modern computers. Boole linked the challenge of implementing the abstract rules of an algebra in concrete physical systems to the question of how systems operating by following rules can make errors. Boole suggested that, if human reasoning consists of processes governed by the rules of an algebra, then reasoning errors might be due to some malfunction in the concrete physical system implementing those rules. This idea resonates with some contemporary accounts of miscomputation, which explain how concrete computing systems can make mistakes in terms of hardware failures (cf., Turing 1950, 449; Fresco & Primiero 2013; Tucker 2018; Colombo 2021).

Technological and conceptual advances in the eighteenth-century textile industry led to complex mechanisms controlled by physically implemented rules in a way that avoided errors (Daston 1994). In the Jacquard loom, different rules govern different sequences of operations for weaving different patterns. Controlling the operations of the loom are punched cards, which are pieces of stiff paper with holes in pre-defined positions. Each pattern of holes implements a rule, thereby constituting an instruction that the machine can physically respond to by performing the appropriate operation. Properly punched and linked together on the loom, these instruction cards store programs for controlling the machine.

Impressed by the workings of the Jacquard loom, the polymath Charles Babbage (1791–1871) designed a programmable computing device called *Analytical Engine,* which also uses punched cards to implement the rules for calculating mathematical functions. Similar to the Jacquard loom, Babbage's

Analytical Engine separates the concrete mechanism performing certain computing operations from the concrete physical support storing the instructions for performing the operations – a functional distinction that would inform the design of digital computers in the twentieth century.

Despite its mathematical capacities, Babbage's computing device was acting in accordance with rules in an apparently mindless way. Ada Lovelace (1815–52), a mathematician and friend of Babbage, explained: 'The Analytical Engine has no pretensions whatever to *originate* anything. It can do whatever we *know how to order it* to perform. It can *follow* analysis; but it has no power of *anticipating* any analytical relations or truths' (Lovelace 1843). This characterisation may suggest that computation cannot be the basis of creativity. More than a hundred years later, the polymath Alan Turing would dub this *Lady Lovelace's Objection* to machine intelligence.

2.2 From Turing to McCulloch and Pitts

Turing (1912–54) was a mathematician, computer scientist, cryptanalyst, and theoretical biologist. He developed a theoretically rigorous, formal model of digital computation now known as the *Turing machine*. He introduced this model in a landmark paper published in 1936 with the title 'On Computable Numbers, with an Application to the Entscheidungsproblem', where he set out to determine which calculations can be performed by mindlessly following antecedently given sets of instructions. Specifically, Turing wanted to answer David Hilbert's *Entscheidungsproblem*, which asks: can an algorithm determine, for each formula in a first-order logical calculus, whether the formula is a theorem? To answer this question, Turing defined simple machines, which are individuated by a finite set of rules for manipulating a finite set of symbols to calculate the values of a mathematical function.

Turing's (1936) paper established three conclusions. First, he argued persuasively that any mathematical function defined over strings of letters from a finite alphabet and such that its values can be computed by following an algorithm (of the sort that a human being can follow) can be computed by some Turing machine. Thus, Turing machines can compute any function that is computable by any algorithm. This conclusion is now known as the *Church–Turing thesis* because, as we will see in a moment, mathematician Alonzo Church proposed an equivalent thesis around the same time.

Second, Turing showed how to encode the rules that define Turing machines in the form of instructions written on Turing machine tapes, which we now call *computer programs*. He also showed how to define special Turing machines, which he called *universal*, to process data in accordance with the programs written on their tape. Given the Church–Turing thesis, this proved that some computing machines can compute any function computable by algorithm so long as they have enough storage space. Today's *digital computers* are more efficient versions of Turing's universal machines.

Third, Turing proved that only a special set of functions defined over strings of letters from a finite alphabet can be computed by following an algorithm, and hence by Turing machines, and hence by digital computers. Most functions defined over a denumerable domain are not computable in this way. Turing also showed that whether a first-order logical formula is a theorem is one such uncomputable function; hence, Hilbert's *Entscheidungsproblem* has a negative answer. While these results provided the mathematical foundation for the design and study of digital computers, they also stimulated new thinking on the significance of computation for explaining cognition.

We should emphasise that 'mechanising' reason by showing that reasoning (logical inference) can be mimicked by Turing machines does not entail that human reasoning is mechanical in the same way. The latter claim, that humans reason in a mechanical way analogous to how Turing machines operate, is much stronger – it was not obviously advanced by Llull, Descartes, Leibniz, Babbage, or even Turing (1936; though Turing 1950 is another matter); Hobbes might be an exception (cf., Isaac 2018a). The first authors who explicitly linked Turing machines to the study of the mind were Warren McCulloch and Walter Pitts (1943) in a paper entitled 'A logical calculus of the ideas immanent in nervous activity', where they 'treat[ed] the brain as a Turing machine' (McCulloch quoted in Jeffress 1951, 32; Copeland & Proudfoot 1996; Piccinini 2004).

McCulloch was a neurophysiologist familiar with the then-recent discovery of the all-or-none law of nervous activity – that is, that typical neurons either send a powerful signal down their axons, which can excite or inhibit other neurons, or do not send any signal at all. Pitts was a young scientist with great mathematical skills, who was working on mathematical models of neural network activity within a mathematical biophysics research group led by Nicolas Rashevsky (1938). Inspired by Leibniz's idea that reasoning can be mechanised via a formal logical system and Turing's way of formalising logical inference via his machines, McCulloch and Pitts joined forces to articulate a theory of how ideas and inferences – what we now call cognition – might be realised in the brain.

In their 1943 paper, they observed that an 'all-or-none' neuronal signal can be represented by a symbol like *1* and the lack of a signal by a *0*. When that is done and appropriate simplifications and idealisations are in place, neuronal circuits can be shown to perform Boolean operations such as AND, OR, and NOT on simple strings of *1*s and *0*s. McCulloch and Pitts showed how any combination of certain Boolean operations can be realised by their neural networks, and they conjectured that the control device of Turing machines can be constructed out of their networks (Kleene 1956 proved their conjecture correct).

In arguing that brains enable thinking by operating like Turing machines, McCulloch and Pitts suggested that viewing the brain as a computing device provides us with a fruitful foundation for studying and understanding brains and their mental functions (McCulloch 1949). The paper of McCulloch and Pitts (1943) impressed several leading thinkers, including John von Neumann and Norbert Wiener.

Von Neumann relied on McCulloch and Pitts's method for designing neural networks to describe the *architecture* (or functional organisation) of a digital computer with a central processor, control and memory units, and input–output mechanisms (von Neumann 1945). He also referred to McCulloch and Pitts's neurocomputational theory of cognition to evaluate the analogy between computers and nervous systems (von Neumann 1958).

The analogy between computing machines and the nervous system became a tenet of the cybernetics movement. Cybernetics emerged from research concerning purposive systems in biology and communication engineering, aimed at understanding and designing self-organising, self-regulating, and self-maintaining systems that rely on feedback signals (Wiener 1948; Ashby 1952). In 1945, one of the originators of cybernetics, Norbert Wiener, wrote to the Mexican physiologist Arturo Rosenblueth about a conference he had recently attended:

> Von Neumann spoke on computing machines and I spoke on communication engineering. The second day [Rafael] Lorente de Nó and McCulloch joined forces for a very convincing presentation of the present status of the problem of the organisation of the brain. In the end we were all convinced that the subject embracing both the engineering and neurology aspects is essentially one, and we should go ahead with plans to embody these ideas in a permanent program of research. (quoted in Abraham 2018, 55)

Soon afterwards, in 1950, Turing suggested that computers will behave intelligently enough to be regularly mistaken for human beings within a few decades (Turing 1950). In responding to *Lady Lovelace's Objection* (1950, 450) that computers simply follow instructions and thus cannot be creative, he replied that, whether through programming or through learning, nothing prevents computers from being creative. Turing's paper opened the door to the ongoing debate about the nature of artificial intelligence (AI) and how it could be reliably recognised.

2.3 From the Cognitive Revolution to Computational Neuroscience

Two ideas from the paper of McCulloch and Pitts (1943) were particularly influential on the emerging cognitive science. One was that Turing machines, and digital computers more generally, provide us with a fruitful model of the mind; this was the basis for what came to be known as *classical AI* and *cognitive science*. The other was that artificial neural networks can implement computations underlying mental capacities; this was the basis for what came to be known as *connectionism* and *computational neuroscience*.

While these two ideas crystallised the insights that mental capacities can be explained by digital computations and that some physical systems can implement such computations, we should not conclude that the history of CTM in the second half of the 1900s coincides with the application of *digital* computation to the mind. Several of the first scientists who considered CTM, including Turing himself (1950, 439), argued that the brain is not digital and may be more similar to an analog computer, where an analog computer is one that can calculate by integrating continuous variables (cf., Gerard 1951; Lashley 1958; von Neumann 1958). As we will see in the following, the controversy about what type of computation best characterises neurocognitive functions continues to this day.

One version of CTM developed in the context of what has been called the *cognitive revolution*. In the 1950s, a new generation of researchers in psychology, linguistics, and computer science agreed on the inadequacy of the behaviourist tenet that behaviour should be explained solely in terms of external stimuli, responses, and learning histories. They argued that adequate psychological explanations can also appeal to mental capacities such as memory, language, and reasoning and should rely on concepts and models from information theory and computer science. As George Miller, one of the pioneers of this 'cognitive revolution', remembers (2003, 142): researchers contributing to the cognitive revolution intentionally avoided referring to subjective conscious experience and 'were still reluctant to use such terms as "mentalism" to describe what was needed, so [they] talked about cognition instead'.

The theoretical prudence recalled by Miller raises the question of how much of the mind can be explained computationally. Strong versions of CTM say that *all* mental capacities, including consciousness, are wholly computational. Weaker versions say that *cognitive* capacities – such as perception, thinking, and motor control – are computational, but consciousness and possibly other aspects of the mind (more on this in the following) are not wholly computational. From now on, we will focus primarily on weaker versions of CTM. We will return to consciousness and other challenges in Section 6.

In the meantime, electronic digital computers became increasingly common both within and outside scientific research in psychology and neuroscience. As they were replacing human computers, this new technology inspired new ways of studying the mind (Gigerenzer & Goldstein 1996a). They afforded a theoretical framework and vocabulary featuring terms such as 'encoding', 'decoding', 'information processing', 'algorithm', and 'heuristic', which the new cognitive scientists could use for finding explanations of cognitive capacities and behaviour. More practically, the electronic digital computer provided cognitive scientists in the 1960s with a new methodology based on computer simulation for formulating and testing theories about cognition (Aaronson et al. 1976).

Philosophers noted these developments. For instance, Hilary Putnam (1960) suggested that the mind–body problem, which concerns the relationship between the mental and the physical, is *analogous* to the problem of explaining the relationship between the abstract, formal states and the concrete, physical states of a Turing machine implemented on suitable hardware. Putnam noticed that the states of Turing machines are individuated in terms of the way they affect and are affected by other Turing machine states, inputs, and outputs. Subsequently, Putnam (1967) argued that mental states can be individuated by the way they affect and are affected by other mental states, stimuli, and behaviours. This way of thinking provides a functionalist solution to the mind–body problem, according to which mental states like beliefs, desires, perceptions, and so forth are defined by what they do, that is, by their causal roles within a system, rather than by what they are made of, that is, in terms of the physical realisers that play the causal roles.

Soon followed by Jerry Fodor (1965, 1968), Daniel Dennett (1969, 1978), and other philosophers interested in the mind–body problem, Putnam's functionalism was qualified in terms of computation: what minds do is to perform computations similarly to Turing machines. Combining functionalism and computationalism ushered in a view now known as *computational functionalism*, namely the view that the mind is the computational organisation of the brain, or, in a slogan, 'the mind is the software of the brain' (Piccinini 2010).

Consistent with the new cognitive science and computational functionalism, psychologists and philosophers started distinguishing different levels at which a computing system can be studied and understood, such as the *knowledge*, *symbol*, and *register–transfer* levels (Newell 1982) or the *computational*, *algorithmic*, and *implementation* levels (Marr & Poggio 1976). It seemed that, similar to computer programs, complex cognitive capacities can be decomposed into simpler functional units, such as sub-routines and production rules, and functionally explained in computational terms.

One reason that computation appeals to cognitive scientists is that computations can manipulate representations in a way that matches their semantic content. For several researchers, it became obvious that the mind was literally software, whose

> ... atoms ... are symbols, which are combinable into larger and more complex associational structures called lists and list structures. The fundamental 'reactions' of mental chemistry employ elementary information processes that operate upon symbols and symbol structures: copying symbols, storing symbols, retrieving symbols, inputting and outputting symbols, and comparing symbols. (Simon 1979, 363)

Combining computational functionalism with symbolic mental representation, the proposal was that an adequate explanation of cognitive capacities requires appealing to syntactic operations defined over language-like symbolic structures and (digital) computational procedures that operate on them. Such mental software was often deemed distinct and autonomous from its physical implementation, so that psychology could and should study mental software independently of neuroscience (Fodor 1975). This view is now known as *Classical CTM*.

The label 'Classical' became widespread in the 1980s as a way of contrasting the symbolic tradition with the *connectionist* tradition. Connectionism goes back to Edward Thorndike (1932), a behaviourist psychologist who offered a general theory of learning. According to Thorndike, learning occurs when organisms connect, or associate, a sensory stimulation with a behavioural response. One of his learning principles was that stimulus-response connections are strengthened by practice and weakened by lack of practise. After the mathematical biophysicist Nicolas Rashevsky initiated the field of mathematical modelling of neural networks and McCulloch and Pitts connected neural networks to Turing machines, another psychologist, Donald Hebb (1949), reformulated connectionism as the idea that learning is the strengthening of connections between assemblies of neurons that act in concert: 'neurons that fire together, wire together'. Soon thereafter, a number of researchers, such as Frank Rosenblatt (1958), put these ideas into practice by designing artificial neural networks that could acquire computational capacities such as image recognition through a learning process that strengthened or weakened the connections between artificial neurons (unlike McCulloch and Pitts's networks, which had a fixed architecture).

According to formulations of CTM grounded in *connectionist modelling*, cognition is explained by the processing of (non-language-like) representations that emerge within neural networks via the learning process. Unlike Classical

CTM, connectionism is explicitly inspired by salient features of biological brains, where neurons are basic processing units taking a weighted sum of inputs from other neurons and transforming such inputs into an output passed on to other neurons. As in biological systems, these transformations can enable artificial neural networks to perform cognitive tasks.

One limitation of early connectionist research such as that of Rosenblatt (1958) was that no one knew how to train artificial neural networks that had more than two layers of neurons, even though it was easy to show that relatively simple computations require at least three layers (Minsky & Papert 1969). This challenge was overcome by new techniques – most prominently, the *backpropagation algorithm* first described by Paul Werbos (1974). This algorithm trains an artificial neural network by using error signals, which quantify the discrepancy between the network's current output and the output the network is *supposed* to return. These error signals are used to adjust the synaptic weights between neurons so as to make it more likely that the network will yield the correct output in the future. The process of adjusting weights begins from the output layer of the network and is then propagated 'backwards' towards the input layer. By updating these weights over many iterations, while the network is fed some input and returns some output, the artificial network can improve its performance, learning a function that maps inputs to the correct outputs.

Boosted by the backpropagation algorithm, by a suite of other procedures for training multi-layered artificial neural networks, and by several examples of how artificial neural networks match empirical data about language, memory, and perception (Rumelhart et al. 1986), *Connectionist CTM* posed several challenges to Classical CTM. Artificial neural networks show more resilience in the face of damage and noisy input than classical models, and the information distributed in their activation patterns and connection weights is sub-symbolic in the sense that it is often difficult to assign a concept (or atomic representational content within a language-like system) to individual components in the network.

Classicists responded to connectionism with objections of their own. A prominent one was that the sort of artificial neural networks that were popular at that time seemed unable to explain seemingly central properties of thought and reasoning like their *systematicity*, whereby a thinker who can think that the dog chases the cat is also able to think that the cat chases the dog (Fodor & Pylyshyn 1988; Marcus 2001; Calvo & Symons 2014). Others criticised connectionism because backpropagation seemed biologically implausible (Crick 1989) or because the amount of training and energy required for an artificial neural network to correctly perform a given task seemed too high compared with human cognition (Lake et al. 2016; Marcus 2018).

In spite of these criticisms, research in neural networks has continued to advance. Since the 2000s, a new generation of systems called *deep neural networks* has achieved extraordinary performance in a variety of cognitive tasks (LeCun, Bengio & Hinton 2015). Deep neural networks include many layers of artificial neurons that implement mathematical operations called *convolution* and *pooling*, which allow them to learn to match or exceed human performance at tasks such as image and speech recognition, language production, and playing games. For instance, there are now deep neural networks that are ranked as the best players in the world at many complex board games, such as chess and Go.

Along with deep neural networks, several other developments have put pressure on a sharp distinction between classicism and connectionism, opening new ways for formulating, understanding, and evaluating CTM. Such developments include hybrid architectures that combine symbolic and non-symbolic components (Graves et al. 2014), an increased blur of the distinction between hardware and software in developing more energy-efficient neural networks (Wright et al. 2022), and a closer interaction within the field of *computational neuroscience* between neuroscience, machine learning, computer science, mathematics, and psychology (Dayan & Abbott 2005; Pouget et al. 2013; Hassabis et al. 2017; Richards et al. 2019). In particular, the rise of computational neuroscience, which developed largely independently of the debate between classicism and connectionism, has re-shaped the discussion of CTM.

Computational neuroscience emerged from combining the attempt to model neurocognitive processes mathematically – the enterprise pioneered by Rashevsky's biophysics group – with the insight of McCulloch and Pitts (1943) that neurocognitive activity processes information by performing computations. These roots intersected with and were subsequently informed by work in neuroscience, such as Hebb's (1949) description of a learning rule for adjusting connection weights in neural networks, Hodgkin and Huxley's (1952) model of neural action potentials, Barlow's (1961) model of efficient sensory coding, and Marr and Poggio's (1976) distinction between different levels of analysis for studying information processing in the brain. The result is a rich field of models of neural computation at different levels of granularity, from dendrites to neurons to circuits to networks.

The term 'computational neuroscience' appeared towards the end of the 1980s. Its coinage is attributed to Eric Schwartz, who organised a conference with the title 'Computational Neuroscience' in 1985. This conference focused on progress in related fields, which were variously referred to as 'neural networks', 'neural modelling', 'brain theory', and 'theoretical neuroscience' (Schwartz 1990). During the same period, the first departments and graduate programs in computational neuroscience were instituted – for example, the

Computational & Neural Systems program at Caltech in the USA started in 1986 – and the first textbooks were published, contributing to crystallising the theoretical and methodological commitments of a new field (Koch & Segev 1998; Churchland & Sejnowski 1992). In the 1990s, an institutionally recognised community of scientists was born. This community included neuroscientists, computer scientists, mathematicians, and physicists who shared a goal, a set of methods, and a foundational assumption. The set of methods included tools from mathematics, computer science, information theory, and physics. The assumption was that 'brains are kinds of computers' (Dayan 1994, 212).

One difference between connectionism and computational neuroscience, according to a common use of these terms, is that connectionist models exhibit cognitive capacities in a way that is neurally inspired but *not* constrained by neuroanatomical and neurophysiological evidence, whereas neurocomputational models aim to include relevant neuroanatomical and neurophysiological details. In recent years, the distinction between connectionism and computational neuroscience has become less significant. Most researchers agree that understanding how computation explains cognition requires close collaboration between psychology, neuroscience, computer science, engineering, and AI (Pouget et al. 2013; Hassabis et al. 2017; Richards et al. 2019; Piccinini & Ritchie forthcoming).

Ongoing debates focus on differences between methodological strategies and styles of explanation, where connectionist and neurocomputational approaches tend to emphasise how cognition emerges from the dynamical interaction of a large number of processing units (McClelland et al. 2010), while other approaches – for example, involving structured probabilistic models of cognition – tend to emphasise optimal solutions to a given problem and then working out how resource-bounded cognitive agents might approximate such solutions (Griffiths et al. 2010).

2.4 Summary

The history we just sketched, from Ramon Llull to computational neuroscience, foregrounds the long and diverse pedigree of key ideas behind contemporary formulations and arguments about CTM. With this historical background in place, we now turn to the questions: what is computation? How should we distinguish between systems that compute and systems that do not?

3 Computing Systems

The term 'computing system' is used differently by different researchers with different disciplinary backgrounds and research goals. Sometimes this diversity causes confusion and misunderstanding in contemporary debates about CTM, especially in an interdisciplinary context. For example, if you use 'computing system' to refer to familiar electronic machines like laptop computers and smartphones, then CTM is false, and no CTM could even be formulated before Electronic Numerical Integrator and Computer, the first electronic programmable computer, was built in 1945.

A formally precise notion of *computing system* emerged at the beginning of the 1900s, grounded in the concepts of *computable function* and *algorithm*. This notion motivates specific and productive questions about the relationship between mind and computing: is it appropriate to develop a CTM without any reference to the material, tangible aspects of a computing system like, for instance, the brain? What kind of computing system is the human nervous system? Is it the same kind of computing system as an octopus' nervous system? Is performing computation the only thing that brains can do? What is an adequate computational explanation for a mental capacity? Should adequate computational explanations of mental capacities refer to mental representations? Can all mental capacities be explained computationally? How can we explain computationally mental capacities like emotion, creative thinking, and consciousness? One can productively clarify and address these questions only after the notions of a *computing system* and a concrete physical system *implementing* computations are made clearer and more precise. This is our task in this and the next section.

3.1 Computable Functions

Before the advent of artificial computers, computers were humans (often women, Light 1999) performing mathematical calculations by using *algorithms* – that is, finite mathematical recipes that can be followed mechanically – for solving mathematical problems of practical interest. What kinds of problems can be solved in this way? And what exactly is an algorithm?

Alan Turing and Alonzo Church answered these questions in the 1930s. They independently proposed precise definitions of the class of functions, defined over a denumerable domain, whose values can be computed by following an algorithm – the so-called *computable functions*. With these definitions in hand, they proved that first-order logic is undecidable – that is, there is no algorithm that returns, for any first-order logical formula, whether that formula is a theorem. So, not all problems can be solved computationally, that is, by following an algorithm for a mathematical function.

Turing and Church's starting point was the notion of a *mathematical function* over a denumerable domain. A denumerable domain is a domain of mathematical objects that can be put into one-to-one correspondence with the natural numbers; examples include the natural numbers themselves and strings of letters from any finite alphabet. A function over a denumerable domain is a mapping that takes in an element of the domain and returns an element of another set called *range*. Squaring a natural number, for example, is a mathematical function from the set of natural numbers to the set of their squares.

To *compute* a mathematical function $f: I \rightarrow O$ is to transform an input i belonging to set I into an output o belonging to set O by following an algorithm. A function is *computable* just in case there is an algorithm that, given an input of the function, returns the corresponding output. For example, squaring a natural number is computable since, given an input i, say 3, there is a procedure consisting in multiplying i by itself, 3×3, which returns the corresponding output 9. Many mathematical questions, including the decision problem that Turing and Church cared about, can be formulated as functions with *yes* or *no* as possible outputs: 'Given a first-order logical formula, is it a theorem?', 'Is the number n prime?', or 'Given an arbitrary polynomial, are its roots integers?' Which of these functions are computable? In other words, which of these functions are such that there is an algorithm for obtaining the correct value for each argument of the function? Since the notion of *computable function* is defined in terms of an *algorithm*, we want to know exactly what an algorithm is; we want a definition of the informal concept of an *effective* or *mechanical* procedure for computing a function that is unambiguous, formally precise, and practically useful.

Turing (1936) developed an abstract, mathematical model of idealised computing systems now known as *Turing machines*. He then argued that his machines could compute any function computable by following an algorithm. Church (1936a, 1936b) defined the computable functions precisely by identifying them with the recursive functions; he also developed a logical system known as *lambda calculus*. Many other models of effective computation, such as *cellular automata* (von Neumann 1951), have been proposed. Each of these models of computation provides us with a precise account of how an output of a function (over a denumerable domain) is computed, given an input. All such models, no matter how different from one another, turn out to be extensionally equivalent, meaning that any function that is computable within one model is computable within any of the others. This confluence offers one reason for the (mathematical) *Church–Turing thesis*, which says that these models capture the class of computable functions correctly.

Let us now zoom in on Turing machines, Church's lambda calculus, and cellular automata. Our aim is to highlight two facts that bear on a correct understanding of the nature of computing systems in a CTM: first, the notion of *computability* can be made precise in seemingly different but mathematically equivalent ways; second, different models of computation capture different aspects of computing systems.

3.2 Turing Machines and Digital Computation

Alan Turing's model of an effective, or mechanical, method for calculating the values of a mathematical function is that of *computability by a Turing machine*. A Turing machine is defined by four ingredients: a finite set of states (one of which is the initial state), a finite set of symbols that can be concatenated into strings, and a finite set of rules governing the operations the machine can perform on symbols and the transitions between different states.

Imagine an unbounded tape, divided into cells containing symbols from an alphabet. Think of this tape as an unbounded memory, where each cell is a memory location for storing a symbol. At any time, a read/write head is positioned over one of the cells on the tape; the head can perform one of the following operations: reading what is in a cell, writing a new symbol, moving the tape left or right by one cell, and halting. You might think of this head as akin to a central processor, which can access one memory location at a time and perform specific transformations on the elements stored in memory. The operations of the read/write head are determined by a set of rules. Each rule calls for determinate actions to be performed on the basis of the current state of the machine and the symbol on the current cell – for example: 'In State n, if the current cell contains symbol x, then write symbol y, move the tape to the right, and transition to State m'.

As an illustration, consider the Turing machine in Figure 1.

In this example, the initial configuration of the machine depends on the initial 'State 0' and the symbol '0' read in the bold cell. Given this configuration, by acting in accordance with the rules, the head writes '1' in the cell it is currently reading, moves the tape to the right, and stays in State 0. Since the head is now reading a '1', it overwrites that symbol with a '0', moves the tape to the right, and stays in State 0. Once again, the head is reading a '1', and so it overwrites that symbol with '0', moves the tape to the right, and stays in State 0. Now the head reads a blank; the writing rule it should follow is 'None', the moving rule is 'None', and the next state the machine transitions to is Halt. This example of a simple Turing machine precisely formalises the notion of an algorithm, or mechanical procedure, for flipping each 0 stored on the tape into a 1 and each 1 into a 0.

Turing machines have several noteworthy features. One is that Turing machines are *digital* computing systems. This means that (i) the values of the variables they operate on are strings of unambiguously distinguishable

State	Symbol read	Write rule	Move rule	Next state
State 0	Blank	None	None	Halt state
State 0	0	Write 1	Move tape to the right	State 0
State 0	1	Write 0	Move tape to the right	State 0

Figure 1 A simple Turing machine that flips each 0 into a 1 and each 1 into a 0

tokens of finitely many types (digits) and (ii) they perform discrete operations on their digits during time intervals of precisely defined duration. Words in a natural language are one example of a string of digits. Although Turing machines offer a precise, formal model of computability and, as we have seen, CTM has been formulated in terms of Turing machines, CTM need *not* be committed to this formulation. The computational theory of mind may or may not posit that the mind operates on digital, let alone language-like, variables or that the mind performs discrete operations during time intervals of precisely defined duration.

Another interesting feature is that the digital variables on which Turing machines are defined and operate are typically complex symbolic representations. Representations are objects that can stand for something else and can be implemented in a concrete computing system – for example, as strings of voltages within memory cells whose activation stands for, say, Greta Thunberg. But again, a CTM may or may not posit that the mind operates on representations, that mental representations are symbolic or digital, or that mental representations must be strings of on or off voltages within memory cells. As we shall discuss shortly, our minds appear to process different kinds of representations from those manipulated by Turing machines.

A third and final important feature is that Turing machines encapsulate some of the basic design principles underlying modern digital, electronic computers – specifically, the separation between memory and processor and the performance of digital operations on digital variables. In addition, *universal* Turing machines – that is, Turing machines that execute instructions encoded on their tape – are a kind of program-controlled, general-purpose digital computing system. Modern digital computers have additional architectural features: their internal memory storage is separate from their input and output devices, their memory registers can be accessed independently of one another, and their processor is much more sophisticated than a Turing machine's read–write head. These additional features make modern computers more efficient than Turing machines, but their computing power remains the same as that of universal Turing machines. But, once again, a CTM may or may not posit that the mind has memory units separate from its processing units, that it stores programs or possesses a central processor or random-access memory registers, or that it consists of general-purpose processors. While some researchers have argued that cognition can only be explained by classical computing mechanisms in the nervous system that implement a read–write memory that is separate from the central processor (Gallistel & King 2010), available empirical evidence from psychology and neuroscience – as we will explain in Section 6 – indicates that many cognitive capacities need *not* be explained by appealing to mechanisms that apply formal rules to symbols stored in a separate read–write memory (see Morgan 2022 for a critical assessment of Gallistel's arguments).

3.3 The Lambda Calculus and Cellular Automata

The lambda calculus is a formal system Church (1936a, 1936b) developed for doing logic by expressing and applying functions. Turing machines offer a *sequential* model of computation, whereas Church's lambda calculus is an example of a *functional* model.

Within the lambda calculus, there are variables that can form expressions, on which we can apply simple cut-and-paste operations to define functions, apply functions to other functions, and resolve functions. A function in the lambda calculus is introduced with the Greek letter λ followed by an input variable, a dot, and some expression corresponding to the output of the function. The 'λ-variable-.' part is the 'head' of a function, and the expression following it is the 'body'. For example, the expression $\lambda x.x$ defines the identity function, which takes x as input and returns x as output, where '$\lambda x.$' is the head and the second occurrence of 'x' is the body.

Resolving a function within the lambda calculus amounts to taking the variable in the head and replacing all of its occurrences within the body with the expression after the function. The variables in the head are those for replacement and are called bound variables. For example, resolving the function

($\lambda x.xy$)(m) amounts to replacing all occurrences of the bound variable 'x' with the expression (m); thus, you cut the expression (m) and paste it into the body, in every place indicated by the bound term 'x' in the head, so as to obtain (my). Once all the lambdas are discharged or there are no expressions after a function, you have resolved the function.

The lambda calculus provides us with rigorous means for defining, applying, and resolving functions. With the lambda calculus, we can define and resolve mathematical functions such as arithmetic operations. We can also define and perform more complex functions for solving tasks like extracting the third element from a list or counting the number of elements in the list.

Cellular automata are the third and final model of (digital) computation we introduce. In this model, several operations are executed concurrently, in parallel, on a finite or denumerable lattice of simple units called *cells*. At any time, each cell instantiates one of several possible, discrete states – say, either the state Green or the state Blue. The state instantiated by a cell at a given time and how this state changes over time depend on a transition rule that is only sensitive to the state of the cell itself and the state of its neighbours. Given a suitable initial distribution of cells, initial configuration of states, and transition rule, cellular automata can mimic any Turing machine and compute the same class of mathematical functions in a purely mechanical way. That is what makes cellular automata a model of (distributed and parallel) digital computation.

Studying self-reproduction was one of the original motivations for developing cellular automata (von Neumann 1966). Over time, interest in cellular automata grew in relation to the concepts of *complexity* and *self-organisation*, as one salient feature of cellular automata is that their cells can evolve into surprising patterns that emerge from the iterative application of simple local rules (Dennett 1991a). The evolution of cells' states in a cellular automaton can thus be interpreted as an algorithmic procedure that, dynamically operating on the cells' states, can give rise to patterns consisting of emergent, complex, and orderly macro-states.

In summary, although the three models of a computing system we have sketched have salient differences, they are equivalent in the sense that any function that is computable by a Turing machine or cellular automaton can be defined and resolved within the lambda calculus and vice versa. Thus, a given digital computing system need *not* be a Turing machine; it can compute in the way of the lambda calculus, cellular automata, or other mathematical models of digital computation.

3.4 Generic Computation and Information Processing

As we have already hinted, not all computation is digital. Before electronic digital computers were invented, there were differential analysers, which performed what were later called analog computations. These analog computers, as they were eventually renamed, need not operate on digital inputs and internal states, need not perform digital operations in discrete steps, and need not return digital outputs. Instead, they can process continuous variables and they can process such variables continuously over time. They still process such variables in accordance with procedures or algorithms, but their algorithms can be defined over continuous variables – they are not limited to functions defined over a denumerable domain like digital computers. Another difference between analog and digital computers is in how they represent their targets (representing a target is not mandatory but typical applications require it). Specifically, since analog computers need not manipulate digits, they need not represent their targets by using digital codes. Instead, analog computers can manipulate analog representations, which represent targets by varying their own values in a way that tracks the values of their targets (Maley 2023).

Besides analog computing, there are a plethora of unconventional models of computation, including quantum computing, reservoir computing, molecular computing, and neuromorphic computing (see, e.g., Adamatzky 2021). All of these depart from classical digital computing in some way or another. To cover all of these types of computing, it is useful to have an umbrella term, *generic computing*, of which the different types are species. What all species of computing have in common is that they operate on some mathematically defined variables and they perform mathematical operations on input variables, internal variables, or both to return output variables that stand in some appropriate relation to the input or internal variables. In other words, all computing is a way to solve general mathematical problems by means of algorithms of some sort (though, possibly, of a sort that a human being cannot follow). CTM, in its most general form, says that the mind is a computing system in the generic sense.

Another common way of formulating CTM appeals to *information processing*. Here, the basic idea is that the mind is an information-processing system. Since there are several notions of *information* (Fresco 2022), we should explicate this notion in sufficiently precise terms if we want to evaluate whether minds or some other concrete physical system processes information and whether their information processing constitutes mental capacities.

We can begin by distinguishing *non-semantic* from *semantic* notions of information. According to one account of non-semantic information, information consists in the value of a degree of freedom of a physical system, where a degree of freedom is a yes–no question required to describe the state of a physical system at a given time (Gabor 1946; MacKay 1969). If a concrete system manipulates a variable with some degree of freedom, then the system processes information in this very broad sense. A related sense of non-semantic information is reduction of uncertainty. In this sense, signals produced by a source are informative to the extent that they reduce uncertainty in a receiver (Shannon 1948). For instance, flipping a coin reduces uncertainty about which face of the coin is up. If a concrete system manipulates signals that decrease the uncertainty of a receiver, then the system processes information. Like the previous one, this notion of information is non-semantic because what signals (or degrees of freedom) stand for is irrelevant to whether and how they count as information.

Many signals reduce uncertainty *about* the state of the source. For instance, light reflected by a flipped coin and travelling from the coin to an observer may reduce the observer's uncertainty about the state of the coin. This is a notion of *natural semantic information* (cf., Dretske 1981). It is semantic because it is *about* the source. It is natural because its semantic content is simply what it correlates with – what it raises the probability of (every information channel is prone to noise, and so the probability is typically less than 1). In contrast, the statement 'the coin landed on heads' carries *non-natural semantic information* about the state of the coin. It is semantic because, again, it is *about* the state of the coin. It is non-natural in the sense that there is no direct correlation or natural process connecting the statement to its semantic value. In fact, the statement may even be used intentionally to say something false.

The notion of information processing is often used interchangeably with that of computation, but the taxonomy we just introduced counsels caution (Piccinini 2015, Chapter 14). Computation surely involves the processing of variables. Insofar as such variables have degrees of freedom or reduce uncertainty, computation involves (non-semantic) information processing. Computation typically involves the processing of variables that carry either natural or non-natural information. Thus, computation typically involves semantic information processing as well. For instance, any computing system that manipulates symbolic or analog representations is an information-processing system in the semantic sense. As we will explain in Section 4, the stronger claim – that computation must always or necessarily involve semantic information – remains controversial.

3.5 From Abstract Computation to Concrete Computing Systems

At this point, someone might think that if CTM is true, the mind must be an abstract entity. After all, the models of computability we sketched earlier were put forward in mathematical logic. According to platonism about mathematics, mathematical entities are abstract objects – that is, objects with no spatiotemporal location and no causal influence on the world. If minds are, say, Turing machines, a platonist might be tempted to conclude that minds are also causally inert abstract objects with no spatiotemporal location.

One problem with this line of thought is that the conclusion that the mind is an abstract entity does not follow from the claim that formal models of computation pick out abstract, mathematical objects. The functions your laptop computer can compute are exactly those functions picked out by the abstract notions of Turing computability and lambda definability. But that does not make your laptop computer an abstract object. Your laptop, although it implements an abstract model of computation, is a concrete object – it is located in space and time, has spatial properties like shape, and has temporal properties like the duration it takes to run a program. You can interact with your laptop, and it can make things happen in the world.

We will not address abstract objects further, except to say that CTM is compatible with any view about abstract objects. If there are abstract computing systems existing outside space–time independently of the beliefs and practices of us humans, then evaluating CTM requires figuring out whether the mind *instantiates* or *exemplifies* an abstract computing system. If there are no abstract computing systems, then evaluating CTM requires figuring out whether the mind satisfies some relevant computational description. On either view, an adequate formulation of CTM requires an account of *physical implementation*, which individuates concrete computing systems, distinguishes them from systems that do not compute, and identifies what function a given physical computing system computes.

Before moving on to existing accounts, it is helpful to point out two other ways of understanding *abstraction* in the context of CTM. The first concerns the meaning, or semantics, of discourse in the computational sciences of brain and cognition. Computational neuroscientists say things like the following: 'the olfactory system computes probability distributions over possible odours'; 'brains use gradient descent for learning'; 'dopamine neurons encode reward prediction errors'; 'Hopfield networks explain associative memory'; 'retinal ganglion cells are linear filters'. Under what conditions are these claims true? Should we take all of them *literally* at face value? Or are they metaphorical, fictional, or otherwise non-literal statements, which should be appropriately reformulated when spelling out their truth conditions? And should these truth conditions refer to any human social practices, intentions, or goals? To answer these types of questions about the semantics of scientific discourse in computational neuroscience, the terms 'abstract' and 'abstraction' are sometimes used to mean roughly non-literal, non-realistic, or fictional.

A second way of understanding *abstraction* concerns scientific modelling. Scientific modelling involves simplification, idealisation, and approximation

(Weisberg 2013; Potochnik 2017; Wilson 2022). In modelling, abstraction consists in stripping away, removing, or ignoring properties from the system being modelled. For example, a model of the linear harmonic oscillator might abstract away frictional forces acting on the oscillator, let alone the colour of the oscillator. The model need not include these details, allowing scientists to focus on a limited number of relevant properties. A computational model of perception is abstract in this sense since it ascribes certain properties to perceptual systems but abstracts away from others. This notion of abstraction as 'stripping away' properties and details from a model is particularly important in relation to our understanding of successful computational modelling practice, to which we will turn our attention in Section 5.

3.6 Summary

A computing system is any system that can yield the values of function $f(x)$ by following an algorithm for transforming x into $f(x)$. If the mind is a computing system, that means the mind can calculate the values of functions in accordance with algorithms and these computations constitute or explain mental capacities. Given a liberal application of this account, one immediate worry is that every physical system might be described as following an algorithm from its initial conditions to its future states and, hence, as computing a function that yields its own states at future times. If so, CTM would be trivially true. If we want to avoid triviality, we should offer an account of how to individuate concrete, physical computing systems, and how to identify what function, if any, a given system computes.

4 Computation in Physical Systems

According to CTM, mental computations constitute or explain mental phenomena and mental capacities. This raises the question of what it means when a physical system performs computations. When can we justifiably say that certain concrete systems like, say, brains, smartphones, and maybe even gene regulatory networks *are* computing systems, but other concrete systems like, say, crystals, livers, tornadoes, and black holes are not computing systems? Without a satisfying answer to this question, we face the worry that CTM is just *trivial* (cf., Sprevak 2018).

One worry is that, if the notion of computation is so weak that *every* physical system *is* a computing system or is fruitfully understood as a computing system, CTM is not very interesting. It provides us with no distinctive insight about minds. One could say, for example, that the tides compute a function from the current location of the Moon to their height; by following an algorithm, they 'decide' whether to rise and by how much. This type of conclusion trivialises CTM at least to a degree. An even more serious worry is that, if the notion of computation is even weaker and every physical system implements *every* computation or at least a large-enough number of non-equivalent computations, then CTM says nothing substantive about minds at all. For if every physical system implements every computation, then every physical system performs *mental* computations; therefore, saying that the mind is a computing system says nothing substantive at all. The view that every physical system is a computing system is known as *limited pancomputationalism*, while the view that every physical system implements every computation is known as *unlimited pancomputationalism*. Accounts of implementation can be assessed based on whether they entail, or avoid, limited or unlimited pancomputationalism.

Accounts of implementation articulate conditions under which a given physical system implements computational processes defined by a mathematical model of computation. In providing us with these conditions, such an account should enable us to distinguish between computing and non-computing physical systems and also between computational and non-computational properties within the same system (Shagrir 2022, Chapter 1). It should also enable us to determine what particular function among several possible ones a given physical system computes when it produces a certain phenomenon (Fresco, Copeland & Wolf 2021).

In evaluating how an account of implementation addresses triviality worries, it is important to clarify why pancomputationalism is worrisome for CTM. First, pancomputationalism violates computational cognitive scientists' and computer scientists' judgements about implementation. Second, at least unlimited pancomputationalism entails that computational explanations of mental capacities are devoid of explanatory power. If computational implementation is trivial, then explaining particular psychological capacities by appealing to the implementation of particular computations will not yield any *distinctive* insight into that capacity since every other physical system, including rocks and rivers and the solar system, would implement every computation. This result would call into question whole research programs in the computational cognitive sciences, where computational explanations of psychological phenomena do seem to enjoy explanatory power and yield a distinctive insight into those phenomena.

4.1 The Simple Mapping Account

According to the simple mapping account of implementation, a physical system computes if there is a structure-preserving mapping between the physical structure of the system at some level of granularity and the formal structure defined by a mathematical model of computation (e.g., a specific Turing machine). If there is such a mapping, then the physical system implements the computation defined by the model and counts as a genuine computing system. Though simple and easy to understand, this account is too liberal since it is too easy to find *some* mapping between the structure of a physical system and that of an algorithmic process.

Hilary Putnam (1988) defends a detailed triviality argument targeting a version of CTM he had previously developed. We encountered Putnam in Section 1 as one of the architects of *computational functionalism* who, he argued, could solve several puzzles about the relationship between mind and brain (Putnam 1967, 1975; see also Block 1978; Fodor 1987). In subsequent works, Putnam came to reject computational functionalism. Putnam (1988, 121–5) argues that it is too easy to map the state transitions between the physical states of a non-cognitive system like a rock onto the computational states defined by an abstract model of computation. If this mapping suffices to implement a computing system, then every physical system implements every (inputless, outputless) computing system.

One natural response to Putnam's triviality argument is to acknowledge that the mapping account is too liberal and add suitable constraints to yield an account of computational implementation that does not trivialise CTM in this way. The constraints can be semantic, counterfactual, causal, or otherwise. Adding these constraints might not only address triviality worries, but – as we will see in a moment – might also result in a more specific version of CTM.

4.2 Restrictive Mapping Accounts

Restrictive mapping accounts attempt to avoid triviality results by restricting acceptable mappings between physical systems and computing systems. Acceptable mappings might be those that capture the causal or dispositional structure of the physical system, support certain counterfactuals, or allow the physical system to be used to predict the evolution of the computing system (Maudlin 1989; Copeland 1996; Klein 2008; Godfrey-Smith 2009; Chalmers 2011; Rescorla 2014; Horsman et al. 2018; Campbell & Yang 2021).

On restricted mapping accounts, no system implements every computation because the physical states triviality arguments appeal to are causally disconnected or fail to support appropriate counterfactuals – for example, counterfactuals of the form, 'If a concrete system were in a physical state p_1, which maps onto the state s_1 of a suitable abstract model of computation, then the concrete system would have gone into a physical state p_2, which maps onto s_2' (Copeland 1996, 341). Even if satisfying this sort of restriction still allows that every system implements some computation, this may not be a serious problem.

Computation may be pervasive in nature, but not everything in nature is an interesting computational system.

Even if these restrictive mapping accounts avoid unlimited pancomputationalism, one may find the conclusion they licence – that everything performs at least some computation – overly inclusive and antithetical to the way computer scientists and cognitive scientists talk about computing systems and distinguish between systems that compute and those that do not. To address this problem, Anderson and Piccinini (forthcoming) propose a 'robust' mapping account of implementation, according to which a physical system P implements a computing system C only if some of P's physical states map onto all of C's states, the physical state transitions between physical states that map onto computational states also map onto computational state transitions, and, most importantly, the physical states that map onto computational states are computationally equivalent to the computational states they map onto, where a physical state is computationally equivalent to a computational state if and only if both states entail the same set of possible computational trajectories. In other words, a hypothetical agent who knew the current physical state and physical dynamics of the system could infer neither more nor less about the computational evolution of the system than an agent who only knew the current computational state of the system and its computational description. Anderson and Piccinini argue that these conditions capture the physical signature of computation and, therefore, should be incorporated into any adequate account of implementation, including semantic and mechanistic accounts.

4.3 Semantic Accounts

According to semantic accounts, a system cannot compute unless at least some of its states possess semantic properties (e.g., meaning, reference, or truth conditions); that is, a system computes only if it manipulates representations (e.g., Fodor 1975; Sprevak 2010; Shagrir 2022). Computing systems differ from non-computing systems because computing systems can manipulate representations or at least they can manipulate appropriate representations in appropriate ways, while non-computing systems cannot.

Semantic accounts rule out rocks and stomachs as genuine computing systems because rocks and stomachs do not manipulate representations. They are also in line with the way many computational cognitive scientists talk about computation and understand the key aims of their field. For instance, in a seminal paper on computational neuroscience, Sejnowski, Koch, and Churchland claim that 'The ultimate aim of computational neuroscience is to explain how electrical and chemical signals are used in the brain to represent and process information' (1988, 1299). Furthermore, appealing to computations over representations appears to carry explanatory power in relation to many psychological phenomena. We will revisit relevant arguments in Section 5. For now, it is just important to notice how semantic accounts can avoid pancomputationalism and ground more specific versions of CTM in semantic properties, such as versions of CTM committed to symbolic computing (e.g., Fodor 1975, 1987; Newell & Simon 1976; Gallistel 1990; Marcus 2001).

Semantic accounts face three problems. First, even assuming that an adequate theory of implementation must posit some semantic constraint, it remains unclear which semantic properties (e.g., reference, intensions, or some normative aspect of semantic content) matter to computation, in which format (e.g., digital, analog, language-like, symbolic, or sub-symbolic), whether such semantic properties involve the system's environment, and where these properties should be instantiated in the system (e.g., in the data, in an algorithm, or both). There is little consensus on how to answer these questions (Ramsey 2016; Egan 2018; Shea 2018).

Second, it remains an ongoing challenge for philosophers and cognitive scientists to explain how mental items such as thoughts, beliefs, and desires can be directed towards, or be about, other specific items – in other words, how mental representations can have intentionality. Note that the concepts of *intentionality* and *representation* are distinct. In fact, the notion of representation can be used as a means to address the problem of intentionality. As Fodor puts it: 'It appears increasingly that the main joint business of the philosophy of language and the philosophy of mind is the problem of representation itself: the metaphysical question of the place of meaning in the world order' (Fodor 1987, xi). This 'doesn't, of course, *solve* the problem of intentionality; it merely replaces it with the *unsolved* problem of representation' (Fodor 1996, 260). While there is little consensus on the place of meaning in the world order, we will discuss some promising approaches in Section 6.4.

Third, it seems that at least some physical computations do not require any appeal to semantics or representations (e.g., Dewhurst 2018). The abstract notion of a computable function we discussed earlier is purely formal and is not defined on representations. So, a semantic constraint on a theory of implementation may generally be unnecessary to distinguish computational from non-computational concrete systems, and what function a given computing system computes (Papayannopoulos, Fresco, & Shagrir 2022).

4.4 Mechanistic Accounts

According to mechanistic accounts, concrete computing systems are mechanisms that perform computations, that is, systems of spatially and temporally organised and causally related components with functions to perform. At least one function of computing mechanisms is that of performing computations (Miłkowski 2013; Fresco 2014; Piccinini 2015; Coelho Mollo 2018).

Physical computation, in turn, may be cashed out as medium-independent manipulation of variables in accordance with a rule (Piccinini 2015, 2020). To illustrate, consider a physical variable solely defined by its degrees of freedom, regardless of how they are physically implemented. This is a multiply realisable variable. Define a mapping from input states of the variable (plus, possibly, internal states of the variable) to output states of the variables – for example, the mapping corresponding to an AND gate. A physical computation, then, is the physical production of such output states from the input states (plus, possibly, internal states) in accordance with the rule defined by the mapping, by a mechanism whose function includes processing such a variable in accordance with such a rule. Since both the mechanism and the variable it manipulates are multiply realisable, the whole process is medium independent. Medium independence entails multiple realisability but not vice versa because a function (e.g., maintaining blood circulation through your body) may be multiply realisable (by different sorts of pumps, whether biological or artificial) even though the medium being manipulated is not multiply realisable (it has to be blood).

Avoiding any direct appeal to semantics and counterfactuals, mechanistic accounts try to avoid triviality worries by individuating concrete computing systems based on their mechanistic properties and particularly on their functions. The reason why your fingernails are not computing systems is that they are not mechanistic structures with the function to compute.

There are at least two options about what determines the computing function of a mechanism. One option is that it is determined by the stable causal contributions that performing computations makes to some goals of organisms; organisms have biological goals, such as survival and inclusive fitness, and may also have other non-biological goals (Maley & Piccinini 2017). Another proposal is that the computing function of a mechanism is determined by the stable causal contributions that performing this function made, in the past, to processes of differential reproduction and differential retention (e.g., processes of evolution, development, or learning) involving organisms with that type of mechanism in a population (Neander 2017). The second option says that a mechanism's computing function depends on the selective history of the mechanism, whereas the first option does not appeal to any historical process but only to how a mechanism's performing computations contributes, now, to the goals of organisms that possess or use that kind of mechanism. Both options share the idea that what fixes the computing function of a mechanism are objective, observer-independent properties of the mechanism.

Mechanistic accounts face various challenges. One challenge concerns the idea of mechanisms possessing the distinctive function to compute. Similar to semantic properties, despite several proposals about what biological functions are and how they should be discovered, it remains contentious whether performing computations is a function of some systems like brains but not of others like stomachs and whether an appeal to biological functions introduces some indeterminacy about a given system's computing function(s).

It is worth noting that mechanistic accounts, and any other non-semantic accounts of computation, still face the question of intentionality and representation. As we shall see in Sections 5.2 and 6.4, there are two ways to handle this. First, there is anti-realism about intentionality/representation, which rejects the idea that intentionality/representation needs to be explained. Second, one can combine a non-semantic account of computation with representationalism about cognition: while computation does not *require* representation, it can occur *with* representation, and cognition is an example of the latter. Of course, the second option faces the same challenges of accounting for intentionality and representation that the semantic account of computation faces.

4.5 Anti-realism about Implementation

According to a completely different approach to triviality worries and the problem of implementation, CTM – however we spell it out – should *not* be understood literally. That is, CTM should *not* be taken at face value as claiming that the mind is literally a computing system or that the brain literally performs computations. Rather, according to anti-realist readings of CTM, some systems like the nervous system are *analogous* to computing systems in some ways and disanalogous in others (Chirimuuta 2021).

Computational approaches to mind, brain, and behaviour offer multiple modelling techniques, whose merits and limitations should be evaluated pragmatically and in a piecemeal fashion in the context of particular models of particular phenomena. Computational modelling provides cognitive scientists with a means for prediction, simplification, and interpretation of behavioural and neural data. But computational models need not be taken literally or as grounding metaphysical claims about the nature and workings of the mind. The value of CTM and of any particular computational model of the mind lies in their *utility* for pursuing certain epistemic or practical goals (Schweizer 2019; Colombo 2021).

Anti-realist accounts of CTM would simply bypass triviality worries by suggesting that, even assuming that rocks and stomachs somehow implement computations, those would not count as legit computing systems because a computational understanding of a rock does not further any scientific purpose. It is pragmatically pointless. Therefore, we should not understand rocks as computing. More generally, one of the selling points of anti-realist interpretations of CTM is that they do not have the metaphysical burden of solving the problem of implementation. Their focus is on how certain target systems are like computers in some ways and how computational modelling is useful for pursuing the wide diversity of goals in computational neuroscience (Körding et al. 2018).

But then, the key challenge for anti-realist accounts is to spell out constraints on computational ascriptions that, though subjective and interest-relative, can make good sense of scientific disagreement about whether a system computes a certain function and also of why computational explanation is apt to yield insight into the behaviour of some systems. Furthermore, compared to a realist understanding of CTM, anti-realist accounts face the challenge of explaining the success of computational modelling of brain, mind, and behaviour. While a literal understanding of CTM would allow us to say that computational modelling of brain, mind, and behaviour is predictively and explanatorily successful *because* the brain *is* a computing system, anti-realists will have to explain this apparent success without appealing to this idea. Finally, insofar as an adequate CTM coheres with the way many scientists think and talk about brains, anti-realist accounts might fail to do justice to this aspect of scientific practice since many cognitive scientists and neuroscientists do take it literally that the brain *is* a computing system. In the words of Christoph Koch's book *Biophysics of Computation*: 'The brain computes! This is accepted as a truism by the majority of neuroscientists engaged in discovering the principles employed in the design and operation of nervous systems' (1999, 1).

4.6 Summary

Different accounts of implementation have been proposed: concrete computation essentially involves suitable mappings, semantic properties, or medium-independent properties and mechanistic functions, respectively. A different approach holds that we should understand what CTM says non-literally in terms of practically useful perspectives, metaphors, or analogies between some aspects of computational models and some aspects of concrete systems studied by cognitive scientists. These different accounts of implementation can be understood as distinct versions of CTM. Even assuming any of these versions successfully addresses the triviality worry that, by itself, would not give us a strong reason to believe our minds are computing systems. Let us now turn to positive arguments in support of CTM.

5 Why Believe CTM?

Arguments that the mind is a computing system typically take the form of an inference to the best explanation for certain mental phenomena, where the theoretical, empirical, and explanatory success of CTM provides us with a (defeasible) reason to believe it. Such arguments rarely purport to show that all aspects of mental life are computational or can be adequately explained computationally; they usually only claim that some important aspects of the mind are at least partly computational. In this section, we will review four types of arguments for CTM, highlighting where these arguments assume different accounts of implementation and how they flesh out CTM.

5.1 CTM Is Needed to Solve the Mind–Body Problem

One reason for CTM is that it helps us solve the mind–body problem. The mind–body problem is how to explain the relationship between thoughts, perceptions, emotions, consciousness, and so on and neurons, muscles, molecules, and so on. We seem to have both physical properties like mass, shape, temperature, and location and mental properties like perceptual experiences, emotions, desires, and beliefs. The mind–body problem is about the relationship between these two kinds of properties and the entities that have them. In particular, do both mental properties and physical properties exist? Are they one and the same? Are they completely distinct? Is one set of properties more fundamental than the other? Do physical states have a causal influence on mental states? Do mental states influence physical states? If they do, how?

There are several views about these questions. According to dualism, the mental and the physical are both fundamental aspects of reality, and they are radically different kinds of things. While dualism has a venerable history and may be intuitively appealing to some, it faces serious problems. One is the problem of interaction, namely, how to account for the apparent interaction between the mental and the physical. Another is the problem of the queerness of the mental, namely, how to avoid making the mental a mysterious, otherworldly 'ghost in a machine'. A third problem for dualism is that it does not cohere with core theoretical and methodological assumptions in the sciences of mind and brain, such as the assumption that mind and brain *are* closely related and that minds are part of the furniture of nature like rocks, trees, and sunshine and should be explainable just like any other natural object.

According to monism, there is only one fundamental kind of thing. This monist position comes in two varieties. According to the idealist variety, everything is mental or depends on a mind for its existence. So, physical properties are fundamentally mental and should be explained in terms of mental properties. Typical idealists see consciousness as the core property of the mind. While some researchers insist that consciousness is both non-physical and more fundamental than the physical world, mainstream philosophers and scientists of the mind endorse the physicalist variety of monism. According to physicalism, everything is physical or depends on the physical. If so, mental properties are ultimately physical and should be explained in terms of physical properties. The main problem for physicalism is that it offers at best an incomplete approach to the mind–body problem. It needs additional claims if it hopes to illuminate the precise nature and workings of the mind.

Physicalists have attempted to address this challenge by trying to explicate the nature of mind in broadly physical terms. Behaviourism, mind–brain identity theory, functionalism, and CTM are examples of how physicalists have articulated this type of explication.

Behaviourism is the view that mental states are dispositions to behave in certain ways. One problem for this view is that it cannot explain occurrent mental states, particularly conscious ones, which do not seem to be reducible to behavioural dispositions. Another problem is that mental states cannot be individually reduced to behavioural dispositions but can only be connected to behaviour as nodes within clusters, which suggests we may never define particular mental states in terms of specific behavioural dispositions.

The mind–brain identity theory identifies kinds of mental states with kinds of brain states. For instance, pain might be C-fibre firing. This view provides a simple physicalist solution to the mind–body problem that coheres with neuroscientific evidence of the intimate connection between mind and brain. It also faces several problems. One is that there are many levels of organisation in the brain, and it is unclear which level is putatively identical to the mind. Another is that there do not seem to be systematic one-to-one mappings between mental state kinds and brain state kinds. Higher level states, such as the perception of a square, appear to be realisable by different kinds of lower level states, such as configurations of neuronal firings. This *multiple realisability* of mental states, *if* it holds (see Polger & Shapiro 2016 and Cao 2022 for critical assessments), entails that mental state kinds are not strictly identical to brain state kinds.

Functionalism is another approach consistent with physicalism. It holds that mental states are individuated by (some of) their causal relationships. For instance, a belief would be individuated by the relationships it bears to certain inputs (e.g., some sensory state), outputs (e.g., some verbal behaviour), and other mental states (e.g., other beliefs, emotions, and desires). What makes a certain physical state mental is what it does, not what it is made of. Since functionalism grounds mental states in their causal relationships, functionalism avoids the problem of interaction that plagues dualism. It also acknowledges, in contrast to behaviourism, that internal mental states are not reducible to behavioural dispositions. And functionalism can also make sense of the multiple realizability of the mental since it individuates mental states not in terms of their physical constitution, that is, what they are made of, but in terms of causal relationships. Causal relations may be realised in different physical structures, just like a mouse trap and a corkscrew can be realised in different physical structures, though they play the same causal role of catching mice and pulling corks from bottlenecks.

Functionalism can be spelled out in various ways, depending on which causal relations are relevant to picking out a certain mental state. According to some early functionalists, the causal relationships that matter for individuating mental states are specified by the laws and generalisations of a scientific psychological theory (e.g., Putnam 1967; Fodor 1968). What sort of theory? A popular suggestion is that the relationships relevant to picking out mental states are specified by a computational theory – a theory that explains cognition via computation (e.g., Block and Fodor 1972; Cummins 1983). The reason for choosing *computational* theories is that computation seems to provide the right sort of explanatory posits. For instance, computation is often seen as a rationally evaluable process over states that can carry semantic content – more similar to mental processes than any other physically realisable processes. So, mental states would be computational states.

In this way, functionalism became intimately associated with CTM, although it should be clear that functionalism does not entail CTM, since functionalism can be true without the functional organisation of the mind being computational, and that CTM may or may not entail functionalism depending on one's view of computation. Still, most accounts of computation see it as a matter of functional relations; if this is correct, CTM entails some version of functionalism. A recent development along these lines is the merging of functionalism and mechanistic explanation, perhaps aided by a mechanistic account of computation (cf., Gillett 2007; Piccinini 2020). According to this mechanistic functionalism, the computational processes posited by functionalism are carried out by computing *mechanisms*, and so CTM provides a computational variety of mechanistic explanation. As functionalism by itself may not go far enough in accounting for the nature and workings of the mind, CTM can helpfully supplement it in a way that makes for the best available solution to the mind–body problem.

Compared to alternative approaches to the mind–body problem, functionalism combined with CTM enjoys a greater degree of explanatory power. It seems more complete as an account of how the mind works. It coheres with the sciences of mind and brain with their emphasis on computational modelling.

It unifies many different mental phenomena displayed by humans and non-humans within a single explanatory framework. It goes at least part of the way towards explaining puzzling mental phenomena such as intelligence and intentionality. Whether functionalism plus CTM can fully account for intentionality and consciousness remains especially controversial; we will discuss these challenges in Sections 6.4 and 6.5, respectively.

5.2 CTM Is the Best Explanation of Cognition

A second reason for CTM is that it provides us with the best explanation of cognition or at least the sort of cognition exhibited by animals with a nervous system. Many animals show exquisite sensitivity to subtle features of their environments, impressive abilities to learn, remember, and solve problems, and amazing flexibility in their adaptive behaviour. It has been experimentally demonstrated that underlying animal cognitions are sensory signals carrying information about the body and environment from sensory organs to the nervous system, where information is processed and motor control signals are sent to motor organs. Mainstream neuroscientists call the information-bearing states that mediate between sensory and motor organs *neural representations*, and the processing of neural representations *neural computation*. Since nervous systems process information by manipulating representations, computing mechanisms are capable of processing information and representations to exhibit cognitive capacities, and no other equally powerful mechanistic explanations of cognition are available, some form of computation over representations is the best explanation of cognition.

Variations of this sort of argument can be found in most corners of the mind sciences. One important variation appeals to computations and some of its properties while eschewing representations. According to some researchers, the notion of representation is too obscure or problematic to underwrite a scientific explanation of cognition. Yet, computation, without involving representations, has the sort of properties, such as flexibility, sensitivity to syntactic structure, or fine-grained control, that are needed to explain cognition (e.g., Stich 1983).

Another version of this inference to the best explanation, popular among philosophers, begins with a characteristic feature of many explanations of mental capacities. Mental states appear to be *causally efficacious* in virtue of how they represent the world to us, that is, in virtue of their semantic content. For example, my desire *to eat a pizza* and my belief *that I can eat a pizza at the pizzeria* will cause me to head to the pizzeria. This explanation seems to work as intended only if my belief and desire are causally efficacious in virtue of their semantic content. Relatedly, mental states can combine inferentially, on the basis of their semantic relationships, to produce other mental states that can stand in logical or semantic relationships that are *rationally evaluable*. For example, if you believe *that greenhouse gases cause global warming* and you also believe *that methane is a greenhouse gas*, then you should believe *that methane causes global warming*. In addition, many kinds of representations are *compositional* in the sense that the semantic content of complex states, like the thought *that methane causes global warming*, is determined by its structure and the semantic content of its constituents. This semantic sensitivity of cognition calls for an explanation. Computation is the sort of process that can be programmed or acquired via training to process representations in a way that respects their semantic properties and relations. Meanwhile, no other causal process is known that can manipulate representations in a way that accords with their semantics. Therefore, some sort of computation seems to be the best explanation of how cognition can respect the semantic properties of representations.

A third apparent cluster of properties of cognition is that mental states are productive and systematic. That cognition is *productive* means that, by combining basic constituents of cognition, one can, in principle, construct an infinite number of cognitive states; for example, one can think *that they are their mother's child*, *their grandmother's grandchild*, or *their great-grandmother's great-grandchild*, and so on. Similarly, cognition seems to be *systematic*: if you have the ability to have certain cognitive states, then you should also have the ability to have other states with semantically related contents. For example, if you can entertain the thought that *Yael loves Ali*, then you should also be able to entertain the thought that *Ali loves Yael*.

To the extent that productivity and systematicity are genuine features of cognition, as opposed to merely apparent, we need to explain why certain cognitive systems display them and how they work. CTM combined with representationalism is the best available explanation for these apparent properties of cognition. So, we have reason to believe that CTM, along with representationalism, is true.

One version of this inference to the best explanation deserves special mention because it purports to favour a specific version of CTM: Classical CTM. According to this argument, the only 'remotely plausible' models of cognition that can explain the causal efficacy of mental states as well as their inferential structure, compositionality, productivity, and systematicity, is Classical CTM, where Classical CTM posits a language-like representational system and computations over such representations driven, in turn, by representations of the rules they follow, analogously to how ordinary digital computers manipulate symbolic structures by executing instructions in their machine language (Fodor 1975, 27, 68–73; Fodor & Pylyshyn 1988; Schneider 2011; Quilty-Dunn, Porot & Mandelbaum 2022). A related point is that classical systems can manipulate symbols with a well-defined syntactic structure, which in turn is closely related to their semantic properties; this means that, 'if you take care of syntax, the semantics will take care of itself' (Haugeland 1985, 106). The conclusion is that we should believe that the mind is a computing device manipulating mental representations with a language-like structure, that is, symbols in a *language of thought* (LOT).

This version of LOT should not be confused with the weaker idea that at least some cognitive states or events have a structure analogous to that of a language. The analogy between thought and language has a long history prior to the invention of digital computers and was eventually combined with CTM as the generic view that at least some mental processes are computations over language-like thoughts (Sellars 1963; Vendler 1972; Harman 1973). This weak version of LOT is compatible with many non-classical computational architectures, including those posited by connectionism and mainstream computational neuroscience. In contrast, Fodor (1975) and other defenders of Classical CTM (e.g., Newell and Simon 1976) posit that cognition is subserved by computing mechanisms much like those of ordinary digital computers.

The core idea of Fodor's account is that a subject's propositional attitudes – beliefs, desires, etc. – should be understood as relations between the subject and sentences in a mental language, which consists in a physically realised, syntactically structured medium of computation that has several properties similar to natural languages and formal logical systems. Mental processes are computational processes consisting of causal chains of rule-governed operations on symbolic, mental representations. According to Fodor (1975), computations over symbols in a LOT provide us with the best explanation of important mental phenomena. Fodor's inference to the best explanation works only if there are no other empirically supported computational architectures that can explain the relevant phenomena as well as or better than Classical CTM. Thus, at best, Fodor's argument is a defeasible reason for Classical CTM.

Consider how to explain the causal efficacy of mental states like beliefs and desires in producing behaviour. These mental states are realised by physical structures that enter certain causal relations with other mental states, perceptual inputs, and behaviour. For example, consider the proposition *that Andrea owns a windmill*. Whether a certain attitude towards that proposition counts as belief (as in, I believe *that Andrea owns a windmill*) or hope (as in, I hope *that Andrea owns a windmill*) depends on the causal role of its realising physical structure. The next step is to distinguish between *content* and *vehicle*. The written sentence 'Andrea owns a windmill' consists of inscriptions on the page with certain shapes and certain spatial relations between them. These inscriptions serve as the *physical vehicle* for the *content* expressed by that sentence. The third, and key, step is to posit that the relationship between the physical structure of the vehicle of a propositional attitude and the formal structure of the content it expresses is the same as the relationship between the structure of a sentence in a language and the structure of the proposition it expresses (where a proposition is, roughly, its semantic content). If the vehicles of propositional attitudes are complex symbolic representations, then one way of understanding the relationship between vehicle and content is in terms of the syntax vs. semantics of a formal system. Sentences in LOT would be purely syntactic objects. The apparent causal efficacy of the content of propositional attitudes in generating behaviour would then be explained by the causal properties of syntactically defined, symbolic vehicles in a LOT.

Furthermore, if the structure of the physical vehicles of representation mirrors the structure of their contents and the computation is programmed accordingly, then the causal transitions between vehicles of propositional attitudes would also reliably track semantic, rationally evaluable, relations between the contents of those attitudes (Newell & Simon 1976; Pylyshyn 1980; Fodor 1987). This is because the LOT would be a *formal system* similar to formal systems in logic such as first-order logic; and we know that, in many formal systems, syntactic derivability tracks semantic validity – and vice versa. Thus, if the vehicles of mental computations are symbolic representations with a language-like syntactic structure, then the operations defined over these symbolic representations can respect semantic and inferential relations holding between their contents. And these semantic and inferential relations 'mimicked' by syntactic relations would explain how transitions between a subject's propositional attitudes can be rationally evaluable as logically valid or invalid.

Finally, consider how Classical CTM could explain alleged properties of thinking such as its compositionality, productivity, and systematicity (Fodor & Pylyshyn 1988; Marcus 2001). If thinking is *compositional*, then the meaning of complex thoughts is determined by the meanings of their constituent thoughts and the rules used to combine them. If LOT includes a primitive base of symbolic variables *and* formal rules for combining those variables into complex expressions on the basis of their syntactic properties, then Classical CTM can explain the compositionality of thought.

Classical CTM could also explain the apparent *productivity* of thinking – that is, the ability to produce an infinite number of new thoughts. If Classical CTM is true and the mind applies syntactic rules to a given finite set of symbolic representations in LOT, and the same rules can be applied again and again to the results of previous applications, then we would have an explanation of how the mind could produce, in principle, an infinite number of new thoughts.

Finally, Classical CTM could explain apparent systematic relations between certain cognitive abilities – for example, your ability to think *that Yael loves Ali* given your ability to think *that Ali loves Yael*. If your ability to think *that Ali loves Yael* consists in your ability to stand in an appropriate relation to the mental sentence ALI LOVES YAEL and you gain that ability by combining the mental words ALI, YAEL, and LOVES, then you can combine those same mental words to form the mental sentence YAEL LOVES ALI, and this gives you the ability to think *that Yael loves Ali*. So, an ability to think *that Ali loves Yael* entails an ability to think *that Yael loves Ali* in virtue of the syntactic and combinatorial properties of LOT.

One problem with this battery of arguments for Classical CTM is that there *are* alternative computational approaches – for instance, connectionist models with the right inductive biases, deep learning architecture, and training – that do not posit a classical architecture and a LOT but might also explain how mental states can be causally efficacious, rationally evaluable, and possess properties like systematicity, productivity, and compositionality to the degree that minds possess them (Churchland 1992; Camp 2007; Calvo & Symons 2014; Pavlick 2022). It is an open, empirical question which particular computational approach best explains cognition. Either way, CTM has proved to be fruitful in clarifying, studying, and explaining many cognitive phenomena.

5.3 CTM Is Supported by the Success of Computational Modelling

A third reason for CTM is that computational models in the mind sciences are successful and they attribute computational properties to the mind; therefore, the mind possesses such computational properties. Many researchers rely on computational modelling to study and explain the behaviour and mental capacities of certain physical systems. Their explanations of how certain aspects of, say, perception, learning, or emotion, are often *successful*, in the sense that they fit a wide range of experimental data, can have broad scope in covering many different phenomena, can generate novel testable predictions, and can be used to answer *what-if-things-had-been-different questions* (Woodward 2003, 6). That is, computational modelling can be used to provide explanations of mental phenomena and capacities because empirically supported computational models not only produce the phenomena and capacities of interest but also enable scientists to identify which features of the brain, body, or environment are important for the occurrence of the *explanandum* and how various changes in relevant and irrelevant features result in changes in the *explanandum*. Thus, successful computational explanations provided by computational models are evidence that the mind is a computing system.

To flesh out and evaluate this argument, we should be clearer on what a computational model is and how computational modelling is successful with respect to good explanation, prediction, and control. In general, scientific modelling consists of constructing representations of some target phenomenon or system in order to make that target easier to study and understand (Hesse 1966; Weisberg 2013; Potochnik 2017). To achieve these aims, scientific models involve idealisation, simplification, and approximation. Simplifications are omissions of known facts about a system. Idealisations are assumptions built into a model that are strictly false but make the model easier to build, understand, and manipulate (e.g., computationally tractable). Approximation is the feature of a model such that it describes a target with less than perfect accuracy; this is a consequence of simplifications and idealisations as well as other factors such as incomplete knowledge. For example, McCulloch and Pitts's (1943) model of neural circuits simplify and idealise neurons by stripping away most of their biophysical properties and by falsely representing them as digital on–off devices. In many cases, idealisations and simplifications do not detract from, but instead contribute to, the success of a model in inspiring experiments, producing clinically relevant insight, or enhancing understanding of a system.

There are many kinds of scientific models, including scale models and ordinary differential equations. The advent of computers provided a new tool in the form of computational models. A computational model is a computer program that starts with a representation of the initial state of a target system and updates that representation by computing subsequent states of the target. In most cases of computational models, no inference is made that the target system is itself computing anything. In the mind sciences, however, computational models are typically put forward as (approximations of) computations attributed to the target system itself. Aside from describing neural computations, computational modelling plays a wide variety of explanatory and practical goals, including inspiring experiments, finding solutions to problems in AI, and developing clinical interventions (McClelland 2009; Körding et al. 2018; Guest & Martin 2021).

Computational modelling in the mind sciences is pitched at multiple scales (or levels of *organisation*) ranging from more microscopic levels, such as molecules, synapses, and cells, to more macroscopic levels, such as brain areas, networks, whole agents, and groups of agents. To clarify their modelling aims at any level where computations occur, many modellers appeal to David

Marr's (1982) account of three levels of *analysis* for a computing system. First, they attempt to describe *what* task the target system is presumed to face; for example, a cash register has the task of adding numbers, or one task of early visual perception is edge detection. Marr calls this, somewhat misleadingly, the 'computational' level of analysis. The task a certain system is presumed to face may be described as an input–output function that the system needs to compute; this need can be motivated in terms of ecological pressures in one's environmental niche (Bechtel & Shagrir 2015). Next comes the *algorithmic level*, where computational modellers study *how* the system works by ascribing specific representations and algorithms to the system, which enable the system to successfully perform the task. Finally comes the *implementation level*, where the aim is to study and understand how physical structures at the relevant scale follow the algorithm and implement the representations.

Different levels of analysis do not compete with one another: they are mutually constraining and mutually informing aspects of one and the same model (Marr & Poggio 1976). To enhance our understanding of any given mental phenomenon or capacity, models are needed at varying degrees of detail and scale, so as to see how the phenomena and mechanisms operating at one scale give rise to those operating at the next higher scale, and so on until we understand how the whole organism behaves. If we could build such an integrated set of models, we would then have a comprehensive, quantitative, and multi-scale understanding of how the nervous system executes which tasks and why (D'Angelo & Jirsa 2022; Wilson 2022).

To appreciate how computational modelling is successful in the mind sciences, it will help to take a closer look at one example. How do honeybees choose which flower to visit next? One answer is that they learn to predict rewards (Montague et al. 1995). This is a kind of *reinforcement learning* (RL) (Kaelbling et al. 1996), whereby agents perform actions and then receive rewards and punishments that they use to select future actions. Following this process enables agents to learn 'how to map situations to actions so as to maximise a numerical reward signal' (Sutton & Barto 2018, 1). That is the 'what' question asked at Marr's computational level of analysis.

A honeybee would maximise reward – that is, getting as much nectar as possible – by gradually learning a mapping between the perceived colour of different flowers and the amount of nectar each flower is likely to yield. That is the 'how' questions at the algorithmic level. As it interacts with the environment, the bee's past experience of what flower leads to what reward is combined with present sensory information to produce a best guess about the yield of different flowers. Based on this guess, the bee chooses which flower to visit; if the amount of nectar received from that flower differs from the expected yield, the algorithm implemented by the bee's brain will compute a *reward prediction error*, which is used to upgrade the honeybee's knowledge base so that its future predictions of reward will be more accurate. Aspects of this algorithmic process are physically realised by a circuit in the bee's brain whose activity is driven by the neuromodulator octopamine.

Honeybees are not the only creatures that rely on RL. Mice, birds, monkeys, humans, and artificial systems that excel at games like Pac-Man and Go implement RL too (Niv & Montague 2009; Mnih et al. 2015; Sutton & Barto 2018). In the last thirty years, RL has become one of the most successful frameworks for studying, modelling, explaining, and understanding mind, brain, and behaviour. RL has stimulated profound developments in computer science and AI (Russell & Norvig 2009, Chapter 21; François-Lavet et al. 2018), as well as in psychology (Dickinson & Balleine 2002; Shah 2012; Sutton & Barto 2018, Chapter 14), neuroscience (Schultz, Dayan, & Montague 1997; Niv 2009; Sutton & Barto 2018, Chapter 15), and psychiatry (Maia & Frank 2011; Montague et al. 2012).

Algorithms and representations that result from RL are widely used to model and study cognitive behaviour in contemporary psychology (e.g., Daw & Frank 2009). Inspired by experimental results about animal conditioning, RL models include representational posits such as 'expectations' and 'reward prediction errors'. These posits have been interpreted in terms of folk-psychological states such as 'belief' and 'desire' (Morillo 1992; Schroeder 2004); they have also been used in cognitive psychology (Dickinson 1985; Kahneman 2011; Evans & Stanovich 2013). One fruitful proposal has been that two different kinds of RL algorithms, viz. 'model-free' and 'model-based' RL algorithms, map onto habits and goal-directed behaviour, respectively (Daw et al. 2005; Dolan & Dayan 2013); more recent proposals blur the difference between habits and goal-directed behaviour in an effort to reflect more accurately the co-operation between distinct processes underlying reward-based learning in humans and other animals (Collins & Cockburn 2020).

With regard to implementation, the dopamine *reward prediction-error* hypothesis has become popular in computational neuroscience (Colombo 2014a). Motivating this hypothesis was the observation that bursts of dopamine activity in certain neural structures reliably occur when the animal receives a reward that exceeds expectations; activity pauses when the reward falls short of expectations; and activity is unchanged when the reward is expected (Houk et al. 1995; Wright, Colombo & Beard 2017). Aiming to explain this pattern of observations, the dopamine *reward prediction-error* hypothesis says that the activity of dopamine neurons in certain regions represents reward prediction errors: 'their outputs appear to code for a deviation or error between the actual reward received and predictions of the time and magnitude of reward' (Schultz, Dayan, & Montague 1997, 1594). Such dopamine neurons would also appear to play distinctive functional roles in the neurobiological mechanisms of choice and learning (Rangel, Camerer & Montague 2008; Richards et al. 2019). This hypothesis has ushered in a flurry of research at the intersection of RL modelling, AI, psychology, and neuroscience that has advanced our understanding of one functional role and information content of some dopamine activity (Langdon et al. 2018; Dabney et al. 2020). That said, dopamine is involved not only in reward prediction errors but also in other functions within other parts of the nervous system.

In clinical practice in psychiatry, the dopamine *reward prediction-error* hypothesis has motivated several computational psychiatrists to rely on RL modelling for studying and understanding psychiatric phenomena. RL models provide psychiatrists with trial-by-trial hypotheses about how a given task may be solved. These hypotheses consist in domain-general algorithmic models, which describe the step-by-step operations (human or animal) participants might carry out to solve a task. By fitting algorithmic models to participants' choice and neural data obtained in the task and evaluating their relative degree of support, computational psychiatrists identify different *computational phenotypes*, viz. measurable behavioural and neural types defined in terms of specific parameters (e.g., temporal discounting or learning rate) extracted from specific computational models of a task (Patzelt et al. 2018; Colombo & Heinz 2019). Computational phenotypes would aid psychiatric classification and might contribute to the development of effective therapies by helping clinicians identify causally relevant (and irrelevant) factors for improving existing learning-based and cognitive-behavioural therapies (Huys et al. 2016; Moutoussis et al. 2019; Colombo 2022).

While computational modelling in the mind sciences raises many philosophical questions about the nature of computational explanation (Kaplan 2011; Chirimuuta 2018; Weiskopf 2018), computational modelling has played productive roles in advancing our understanding of mind, brain, and behaviour. Ascribing algorithmic processes and representations to target systems of interest at multiple scales – from molecules to cultures – helps scientists make precise predictions about the system's behaviour, interpret experimental data in a theoretically grounded way, unify in a single encompassing computational framework like RL a range of phenomena and observations, identify promising interventions on the system to achieve certain desired effects, and bridge biological intelligence and AI.

RL modelling, similarly to many other approaches in contemporary computational cognitive science and neuroscience, does not merely consist in mathematical models and computer simulations of target neural or behavioural phenomena.

In contrast with, say, climate models, RL modelling and models in computational neuroscience more generally allow researchers to formulate precise hypotheses about what a certain neural circuit computes. While the generic claim that brains are computing systems does not offer a strong guide to research, those more specific computational hypotheses do, whether they appeal to RL or some other computational framework. In fact, many hypotheses about which function a target neural structure computes and how it computes it are supported by empirical evidence. Some researchers have even modelled large portions of a brain in an attempt to show how brains produce behaviour, aiming, eventually, to simulate an entire brain with all of its cognitive capacities (Markram 2006; Eliasmith et al. 2012; Colombo 2017). While the empirical success of specific computational hypotheses does not imply that everything that the nervous system does is computing, it does indicate that at least one important thing the nervous system does is computing, which gives us reason to believe CTM.

5.4 CTM Explains the Success of Artificial Intelligence

The field of AI builds, studies, and tries to understand non-biological intelligent systems (Simon 1969; Russell & Norvig 2009). Researchers in AI have successfully pursued a variety of goals, including the design of systems that mimic human intelligence (McCarthy 1959; Lake et al. 2016), the development of algorithms for capacities such as learning, pattern recognition, causal inference, problem solving, and language processing (Pearl 2000; Spirtes et al. 2000; Bishop 2006; Goodfellow, Bengio & Courville 2016), and solving problems we often encounter in our everyday life by engineering devices such as virtual assistants, self-driving cars, brain–computer interfaces, online dating apps, news-ranking systems, credit-scoring algorithms, and autonomous weapons (O'Neil 2016; Murphy 2019; Rahwan et al. 2019).

Typically, AI researchers are not overly concerned with how similar AI systems are to biological systems. Yet, AI and CTM are intimately related. To the extent that CTM is correct in asserting that minds are computing systems, it should be possible to reproduce mental computations, or at least computations equivalent to mental ones, in artificial systems. And to the extent that AI makes progress in building computing systems that exhibit mental capacities, it lends plausibility to CTM.

As you may recall, Turing (1950) addressed the question, 'Can a machine think?' He anticipated the worry that non-biological systems, no matter how sophisticated, would at best *simulate* thought without genuinely implementing it (cf., a computer simulation of a tornado cannot uproot your house). We will return to this worry in the next section. For now, suffice it to highlight that Turing reformulated that question by asking, 'Can a machine be linguistically indistinguishable from a human?' He proposed a scenario, now known as the *Turing test*, where a human judge evaluates whether an agent sequestered in a room who is answering the judge's questions is a human or an artificial computer. The computer passes the Turing test if the judge cannot tell which interlocutor is human and which is a machine.

Though controversial, the Turing test continues to inform discussions of the foundations of AI; for better or worse, several researchers use it as a benchmark to assess the possibility of genuine AI. For example, in 2022 and early 2023, a Google chatbot called LaMDA (Language Model for Dialog Applications) and an OpenAI chatbot called ChatGPT (Chat Generative Pre-trained Transformer) showed such sophisticated linguistic capacities that some observers claimed they are sentient (cf., Wiese forthcoming). Most researchers replied that mimicking human conversation does not show that a machine is sentient. Still, as Turing already pointed out, sentience need not be a necessary condition for intelligence, and LaMDA and ChatGPT surely exhibit linguistic behaviour that would require a great deal of intelligence in human beings. While more recent benchmarks for AI have problems and are becoming increasingly easy to meet, which raises the question of what they actually measure (Kiela et al. 2021), the important point for our purposes is that current technologies suggest AI is possible and, therefore, there is something right about CTM.

Work in AI can contribute to how we should understand CTM and vindicate its key claim that minds are computing systems of one type or another. For example, contemporary researchers working in a tradition allied with Classical CTM claim that manipulating 'symbolic' (i.e., language-like) representations based on hard-coded algebraic rules is essential to achieving genuine AI since it would be essential to general intelligence and common sense reasoning in humans (Tenenbaum et al. 2011; Marcus & Davis 2019). Other researchers contend that biologically plausible intelligence need *not* rely on built-in symbolic representations and algebraic rules for manipulating symbols. Adaptive agents can acquire the ability to *use* symbols in virtue of basic capacities for pattern recognition and completion from repeated interaction with the environment (Brooks 1991; Clark 1993; Bengio, Lecun & Hinton 2021; Santoro et al. 2021).

In any case, a better computational understanding of the workings of biological brains has inspired the design of powerful AI systems like deep neural networks that can perform tasks such as playing games at superhuman levels (Hassabis et al. 2017). In turn, similarly to what has been happening in the field of RL we discussed in Section 5.3, algorithmic models developed in AI, which specify an agent's functions, learning rules, and architecture, are providing brain scientists with precise frameworks for studying and explaining how brains of biological agents might perform perceptual, cognitive, and motor tasks

(Richards et al. 2019; Lillicrap et al. 2020). The common ground in this cross-pollination between AI and the mind sciences is the idea that minds are computing systems; researchers working at this intersection share the language of computation, information-processing, and computational modelling to successfully achieve a variety of theoretical and empirical goals (cf., Bowers et al. 2022; Buckner forthcoming).

The idea of building an artificial mind is clearly a sensitive topic, prone to hype and fraught with ethical implications. A couple of popular claims are that soon we will be able to 'upload' our minds onto computers, thereby achieving 'digital immortality', and we will reach a 'singularity' whereby computers will surpass human intelligence and begin to accelerate their progress without human help. Neither of these provocative claims is especially well supported by the available evidence. As to mind uploading, brains are probably too complex and delicate for us to be able to simulate an individual's brain down to the details of their personality (Mandelbaum 2022); in addition, there is no compelling reason to believe that such a simulation would be conscious; in any case, such a simulation would be a copy of the individual, not the individual themselves (Corabi & Schneider 2012; Piccinini 2021). As to the singularity, we do not understand general intelligence well enough, or how to reproduce it in machines, to be confident about whether and when artificial general intelligence will be achieved and what this will mean for the evolution of AI (for a sceptical take, see Landgrebe and Smith 2022). Hype aside, to the extent that work in AI advances our understanding of the nature of mind and AI systems make rapid progress in successfully tackling a great variety of cognitive tasks, AI supports the plausibility of CTM.

5.5 Summary

This section has reviewed four arguments in support of CTM. First, CTM helps solve the mind–body problem. Second, it is the best explanation of cognition. Third, it is supported by the empirical and theoretical success of computational modelling in the mind sciences. Fourth, it explains the success of AI and is empirically vindicated by the extraordinary performance of AI systems on many cognitive tasks. In unpacking these arguments, we emphasised that sometimes they make specific assumptions about how CTM should be understood. Our next steps are to evaluate some of the assumptions of these specific versions of CTM while reviewing some prominent arguments against CTM.

6 Challenges to CTM

Arguments against CTM take one of two forms. According to the first set of objections, mental abilities are best explained non-computationally because minds are more powerful than computing systems, minds are embodied and environmentally embedded, or brains do not work like computers. According to the second class of objections, CTM is insufficient for explaining some properties of the mind, such as intentionality or consciousness. In this section, we will review both classes of challenges and highlight how, similarly to arguments in favour of CTM, different objections make different assumptions about the scope and nature of CTM. Objections to CTM have stimulated fruitful debates, which have helped clarify the nature of neural computation and the scope and aims of computational modelling in the mind sciences. At best, however, challenges to CTM show that minds are not *only* computing systems – something else besides computation may be needed to explain some properties of minds and brains. To see why, let us begin by examining how resources constrain computation.

6.1 Tractability Constraints

Minds, brains, and artificial computing systems have limited resources. Consider a cognitive capacity such as deciding whether to flee a potential predator. You might think that, once that capacity is defined as the computation of a function, how long it takes to compute that function is irrelevant to CTM. You would be wrong. After all, whether an agent takes 1 second or 1 minute to start running away from a predator is a matter of life and death. And it is not just time that matters for an agent's fitness. Energy expenditure is also crucial. Since biological computing systems consume energy and operate under thermodynamic and metabolic constraints, they must be efficient if they are to stick around and generate offspring.

In Section 3, we qualified digital versions of CTM in terms of effective computability, where a function is effectively computable if and only if there is an algorithm that will transform every input of the target function into the appropriate output. As you will recall, this notion of effective computability has been formalised variously by different mathematicians and computer scientists (e.g., Church 1936a, 1936b; Kleene 1936; Turing 1936). But some computable functions take enormous amounts of time and resources to compute – amounts of resources that are not available to organisms in the wild. Therefore, if the functions characterising mental capacities are computable by Turing machines, they must belong to a *proper subset* of the functions that a Turing machine can, in principle, compute (Garey & Johnson 1979). Which subset? At least the subset of *tractable* functions, where the notion of computational tractability can be made precise with the tools of *computational complexity theory*, which is the study of how resources (e.g., time, memory space, etc.) required to solve a problem (i.e., to compute a function) scale with the problem size (Aaronson 2013; Garey & Johnson 1979).

Tractability results from computational complexity theory are formulated in terms of functions defined over a denumerable domain. As we pointed out earlier, some versions of CTM are formulated in terms of other notions of computation, such as analog or neural computation, where computational complexity theory may or may not directly apply. Nevertheless, tractability considerations analogous to those of computational complexity theory – that is, considerations about the amount of time and resources needed for computing a certain function – apply to any notion of computation. Such considerations about computational constraints bear on CTM in at least three ways.

First, they ground a 'tractable cognition thesis' (van Rooij 2008), according to which the functions posited by CTM to explain cognitive capacities should be restricted to a subset of those that are effectively computable in a realistic amount of time and with the use of a realistic amount of energy and other limited resources.

Second, tractability constraints motivate cognitive scientists dealing with apparently intractable cognitive functions to investigate approximate algorithms or heuristics. A heuristic procedure is one that is not guaranteed to find the optimal or correct solution to a problem every time; yet, it will find either a sufficiently good solution or an approximation of the correct solution most of the time. For example, the computing processes posited by many Bayesian models of cognition are intractable – they cannot be performed by resource-bounded systems like our brains in a realistic amount of time. Chater et al. suggest that approximate algorithms may address this intractability of Bayesian computations: 'full Bayesian computations are intractable ... the fields of machine learning, artificial intelligence, statistics, informational theory and control theory can be viewed as rich sources of hypotheses concerning tractable, approximate algorithms that might underlie probabilistic cognition' (2006, 290). One problem with this approach is that approximate algorithms or heuristics *cannot* compute computationally intractable functions such as those posited by Bayesian models. What they *can* compute is alternative functions that are tractably computable. Therefore, those alternative, tractably computable functions are the ones that should provide computational theories of cognitive functions (van Rooij, Wright & Wareham 2012). An approach to tractability consistent with this observation is to leverage ecological, biological, and physical constraints on computing to refine the characterisation of a target mental capacity. Specifically, one way that mental capacities might plausibly satisfy the tractable cognition constraint is by tapping into resources afforded by their embodiment and environmental embeddedness; for example, by following 'ecologically rational, fast, and frugal' heuristics that lead to good decisions in many cases (Gigerenzer & Goldstein 1996b; Lieder & Griffiths 2020). One may also identify features of actual biological brains, like their slow, noisy, and imprecise style of computing, which allow them to compute at relatively low metabolic costs (Sterling & Laughlin 2015; more on neural computing in Section 6.3).

Third, tractability constraints on CTM raise the question of whether some mental capacities are simply computationally intractable and there is no way around it. For example, mathematics is *undecidable*, meaning that there is no algorithm for determining whether any given mathematical formula is provable within a sufficiently powerful formal system (Turing 1936). Yet, Turing (1950) points out that mathematicians can invent new methods of proof, thereby proving more and more theorems over time. This might suggest that mathematicians have a mental capacity that outstrips that of any computing system. A related point is that any sufficiently powerful and consistent formal system is *incomplete*, meaning that there are true formulas that are not provable within the system, including the formula that states that the system is free of contradictions (Gödel 1931). Yet, by reflecting on the properties of such formal systems, mathematicians can demonstrate that such systems are incomplete. Again, this might suggest that mathematicians have a mental capacity that outstrips that of any ordinary computing system (Gödel 1951; Lucas 1961; Penrose 1989). If that were correct, the mind – or at least some mental capacities – would be more powerful than any ordinary computing system, and general AI would be impossible (using standard computing technology). The mind would have to be either something other than a computing system or a 'hyper-computer', a computing system more powerful than any Turing machine (Copeland 2000).

Turing labelled the above the *mathematical objection* to machine intelligence. He observed that the objection would go through only if mathematicians were infallible at finding new methods of proof or establishing the consistency of formal systems. In other words, the objection would go through only if mathematicians had these capacities with the sort of reliability that an algorithm has. But human mathematicians make mistakes, and it might take them lots of trials and errors before they prove new theorems correctly. If (ordinary) computing systems are allowed the same latitude – for instance, by altering their own heuristics in ways that are partially random and hence uncomputable – they might find ways to match the discoveries of mathematicians (Turing 1950; Piccinini 2020, section 11.2.5).

In conclusion, computation is bounded by resources, and this constrains the search for plausible computational explanations of mental capacities. As we have hinted, one resource for computing efficiently may be the embodiment and embeddedness of the mind – its situatedness. But some researchers present situatedness as a reason *against* CTM.

6.2 Situatedness and Dynamics

Situated cognition is a heterogeneous research program that emphasises the importance of the body and environment to cognition (Gallagher 2006; Shapiro 2010). While neural systems play an important role in mental capacities, advocates of situated cognitive science have suggested that the brain, body, and environment must be studied as a dynamic whole (Clark 1998). According to this situated approach, cognition is essentially *embodied, embedded, enactive, extended*, and imbued with *affect* (4EA cognition). Cognition is embodied if the body plays essential roles, embedded if the environment plays essential roles, enactive if action plays essential roles, extended if it takes place to some extent within the body and environment, and affective if affect plays essential roles. For example, cognitive processes might *extend* beyond the brain into the body and environment, as when we navigate the Internet, gather information, and then store the results in our smartphone. A related point is that cognition is essentially dynamical, meaning that action unfolds in real time within a changing environment, and the brain, body, and environment are dynamically coupled so tightly that some phenomena cannot be understood if we only focus on the activities of the brain and consider the environment and body merely as sources of inputs and recipients of outputs.

The situatedness of cognition is sometimes pitted against CTM by objecting that CTM ignores that the mind is essentially embodied, embedded, enactive, affective, and dynamical. For example, Turing (1950, 434) seems to suggest that a machine can be deemed intelligent whether or not it has 'artificial flesh' or even legs and eyes. In contrast, some proponents of embodied cognition claim that '[e]mbodied cognition . . . sets itself in clear opposition to what it sees as the prevailing stance in cognitive science and psychology, that is, cognitivism and computational methods that abstract mental performance from the full functioning of the body in its environment' (Varela, Thompson, & Rosch 2016, xlvii). If CTM ignores the situatedness of mind, then the best explanation of mental capacities cannot be computational: 'The mathematical methods of nonlinear dynamical systems theory, employing differential equations rather than computation [would be] the primary explanatory tool . . . for explaining cognition as interaction with the environment' (van Gelder 1995; Silberstein & Chemero 2012, 38).

In response to this *situatedness and dynamics objection*, there are three points to make. First, the objection usually assumes a specific version of CTM, namely Classical CTM, where computers are understood as digital machines that manipulate symbolic representations by executing programs. Embodied cognitive scientists reply that real-time perception, action, learning, and reasoning in the real world are *not* best explained by digital algorithmic manipulation of language-like vehicles inside the agent. For example, according to Hubert Dreyfus (2002a, 2002b), 'skilful actions' such as people in an elevator shuffling until they are at an appropriate distance from each other do not involve language-like representation but rather a non-representational, spontaneous responsiveness to the demands of a situation that is best explained in terms of dynamic causal couplings among brain, body, and environment. Dreyfus supports this claim by combining analyses and arguments from Merleau-Ponty's *Phenomenology of Perception* with Freeman's (1991) connectionist and dynamical system models of brain function. Once CTM is considered in its full scope, however, attractor states in dynamical systems and weight spaces of connectionist networks can still be understood as (non-classical) kinds of internal representations and their processing as (non-classical) kinds of computations, which may well play an explanatory role in accounting for some types of actions such as following social norms (Colombo 2014b). Dreyfus's appeal to connectionist models and dynamical systems theory indicates that his target is Classical CTM and, in particular, Fodor's LOT. As should be clear by now, however, CTM need not be classical, language-like representation is only one of several possible formats mental representations might take, and, in any case, there are even versions of CTM that reject representationalism by holding that, while computation explains cognition, the computational vehicles do not represent anything (e.g., Stich 1983).

Second, the situatedness and dynamics objection sets up a false dichotomy between computational and dynamical modelling. Minds and mental phenomena obviously take place in time, occupying a certain spatiotemporal region. Both brains and digital computers are literally physical dynamical systems since they change over time as a function of their current states; and neural and psychological dynamics can helpfully be modelled at multiple temporal and spatial scales using the mathematics of dynamical system theory (Beer 2000; Smith & Thelen 2003; Izhikevich 2007). Far from computation and dynamics being mutually exclusive explanatory approaches to mind and brain, they are complementary, mutually compatible approaches (Eliasmith 1996; Clark 2013; Beer & Williams 2015).

Third, the situatedness and dynamics objection obfuscates the crucial roles of idealisation, simplification, and the diversity of scientific aims in computational modelling of the mind. While some computational models of mind and brain neglect time for some purposes, many computational models incorporate dynamical and temporal aspects of mental capacities and brain functions. Even advocates of Classical CTM are often concerned with time and dynamics (e.g., Newell 1990, sections 3.3 and 3.6). And typical models in computational neuroscience, such as Hopfield's (1982) model of addressable memory, consist of sets of differential equations, which define dynamical systems. Even though Turing seems to downplay the importance of embodiment, *some* computing systems are embodied, embedded, extended, and can exhibit what are known in the *enactivist* tradition as *autonomy*, in virtue of which a system can generate and maintain itself amid external perturbation, and *functional closure*, where the system's outputs loop back through its environment and constitute its next input (Maley and Piccinini 2016; Villalobos & Dewhurst 2017, 2018). Thus, computing systems can display properties key to situated cognition. The point is that any physical system can have multiple quantitative models, and which model and idealisations might be best to capture a given mental phenomenon will partly depend on specific modellers' aims in their research context (Eliasmith 2009; Weinberger & Allen 2022).

In conclusion, computation can be disembodied and atemporal, but it need not be. Whether a target phenomenon is best modelled in terms of differential equations will depend on relevant empirical data about the phenomenon, available technologies for investigation, and the particular epistemic or practical concerns of the modeller (Clark 1997). So, there is no conflict between CTM, situatedness, and dynamics (Colombo 2009; Miłkowski 2017; Isaac 2018b). CTM is compatible with cognition being embodied, embedded, enactive, extended, and affective (Wilson 1994; Clark 2008). That said, is computation a helpful way to understand how the brain works?

6.3 Brain Function

One implication of CTM is that the nervous system, and the brain in particular, is a computing system. But some cognitive scientists and philosophers object that the nervous system, and the brain in particular, is *not* a computing system and terms such as 'information processing', 'coding', 'computing', and 'algorithm' are misleading metaphors when employed to explain mental capacities. If this is correct, then CTM is false.

Specifically, some suggest that psychological properties belong to whole animals and it is a mereological fallacy to attribute them to brains or brain parts (Bennett & Hacker 2022). Others point out that digital computers are serial machines whose components are either on or off, but nervous systems are not digital in this way, can perform many tasks in parallel, and lack a processor separate from memory components (Edelman 1992; Globus 1992; Spivey 2007). Yet others argue that perceptual systems do not process or encode information and do not rely on any kind of representation at all; instead, given the real-time challenges they face in their ecological niches, perception directly picks up affordances from the environment (Gibson 1979; Chemero 2011; Brette 2019). Finally, some argue that mental capacities are not medium-independent but distinctively influenced by their underlying material substrate, which would at least call into question the multiple realisability of computational structures (Bell 1999; Cao 2022). Let us examine each of these criticisms.

The claim that ascribing psychological properties to brains is a mereological fallacy is itself a fallacy of attempting to legislate language without considering how people actually communicate. Correct language use is not fixed; it evolves as science evolves (Wilson 2006). In addition, both laypeople and scientists ascribe properties to both a system and its parts when they identify the parts that play the main role in the relevant phenomenon. For example, we say that stomachs (as well as people) digest food or that 'the mouth speaks',[2] and we commit no mereological fallacy in doing so. By the same token, there is nothing wrong with explaining how people perceive, think, or make decisions by invoking the cognitive capacities of their brains. When psychological properties are ascribed to brains or artificial computing systems in reasonable ways, they do not hinder communication and inquiry. Instead, they generate testable computational hypotheses, models, and explanations (Jonas & Körding 2017; Figdor 2018).

Evidence that brains are not digital, serial computing systems of the same kind as your phone or laptop might carry some weight against Classical CTM, but it is a red herring for evaluating CTM in general. It is more and more widely recognised that neural computation is different from computation within the artificial computers we are familiar with; this only implies that brains are not *that* kind of computer.

Since the late nineteenth century, with the aid of new technologies and methods for data analysis of increased sophistication and reliability, evidence has accumulated concerning the anatomy and physiology of the nervous system of several organisms at multiple spatial and temporal scales. Nervous systems include brains, which divide into anatomically distinct yet highly interconnected sub-systems and areas. Such sub-systems and areas form large-scale networks that work synchronously to solve tasks and control behaviour (Sporns 2016). In turn, sub-systems and areas consist of dynamic biological networks made of several types of neural and glial cells. Neural systems are biologically robust (Marder & Goaillard 2006) and show both functional degeneracy and segregation (Price & Friston 2002; Sporns 2016). Neurons can ask and receive energy as needed and are very energy efficient; their causal interactions depend

---

[2] Cf. Luke 6:45: 'A good person out of the store of goodness in his heart produces good, but an evil person out of a store of evil produces evil; for from the fullness of the heart the mouth speaks'.

on distributed, parallel patterns of noisy, slow, and redundant electrochemical activity (Laughlin et al 1998; Montague 2007, Chapter 2; Faisal, Selen & Wolpert 2008; Sterling & Laughlin 2015). In contrast, artificial digital computing systems are fast and precise, yet brittle and energetically wasteful.

If you want to understand how your laptop works, it is crucial to know that it encodes information as strings of bits. A bit is a unit of information with one of two possible discrete values, commonly represented as '0' or '1', which can be physically implemented in a laptop computer by two distinct electrical voltages. Understanding this digital code enables you to understand how a laptop performs computations. The same applies to the brain. If you want to understand how the brain works, you should try to identify the neural code, how the neural code enables neural representations, and how neural computations process neural representations. In the next section, we will discuss whether and how neural representations explain intentionality; here, we focus on whether there are neural codes and representations in the first place and what their structural properties are.

Though some researchers believe it is misguided to say that there is a neural code (Bickhard and Terveen 1995; Brette 2019), neural coding is a central topic of research in computational neuroscience and there are several hypotheses concerning the way neural activity carries information based on different coding schemes (Knill & Pouget 2004; Dayan & Abbott 2005, Part I; Quiroga & Panzeri 2013). One of the most fundamental empirical observations in this area is that typical neurons send signals – that is, action potentials – at a rate, or frequency, that varies monotonically with the intensity of their excitatory input (Adrian & Zotterman 1926). This correlation between intensity of the stimulus and rate of neural signals is called *rate coding*. Another plausible way that some neural signals encode information is by their precise timing; this is known as *temporal coding*.

To identify the neural code in greater detail, computational neuroscientists rely on recordings of action potentials from individual neurons and populations of neurons in animals presented with perceptual stimuli, such as different odours or edges with different orientations, or performing an experimental task, such as finding their way out of a maze or choosing between different actions to obtain a reward. Given a pattern of neural activity that maps onto a certain stimulus or task, computational neuroscientists often identify its causal role in a mechanism underlying a certain behaviour or mental capacity via experimental interventions and computational modelling. When neural activity encodes external variables, can occur in the absence of the target variables when needed, and plays a role in guiding behaviour, neuroscientists call it *neural representation* (e.g., Baker et al. 2022; Piccinini 2020, Chapter 12; Poldrack 2021), which are a species of what philosophers call *structural* representations

(Gładziejewski and Miłkowski 2017; Lee 2018). Sometimes neuroscientists also bring evolutionary, developmental, or adaptive considerations to bear on what a pattern of neural activity represents (Cao 2018; Cisek 2019). Working out systematic mappings between stimuli, tasks, and neural activity at multiple spatial and temporal scales, coupled with adaptive considerations, contributes to cracking neural codes, showing how neural codes enable neural representation, and understanding how neural computations over representations produce, regulate, and control behaviour and mental capacities.

One question that bears on the nature and plausibility of different versions of CTM is whether neural computation is digital, analog, or *sui generis*. To answer this question, let us illustrate the notions of *digital* and *analog* by means of simple examples. One example of an analog measuring device is a mercury thermometer. The mercury in this device represents the variable temperature by co-varying with it: as ambient temperature increases, so does the level of mercury in the thermometer. Another analog measuring device is a sundial, where a rod casts a shadow onto a platform indicating different times. As the sun changes its position, the rod's shadow changes as well, thus representing the change in time. In both systems, there is systematic co-variation between what they represent and how they represent; this co-variation, at least in these examples, involves continuous variables.

Analog computers can be understood as computing systems that can process continuous variables according to rules or, when representation is involved, as systems that manipulate analog representations, which represent their targets by systematically co-varying with them (Maley 2023). Co-variation in analog computers can involve continuous variables, like the actual voltage level of a unit in a circuit at a certain time; it may also involve discrete variables, like the 'on' or 'off' state of a unit in a circuit.

Digital computers, instead, can be understood as systems that process strings of digits in accordance with rules or, when representation is involved, as systems that manipulate digital representations, such as strings of bits that carry information about a target. Unlike analog representations, digital representations need not systematically co-vary with what they represent – for example, a digital representation in base-2 of the variable temperature does not literally increase in any physical sense, like a mercury column does as temperature increases.

Are neural representations analog, digital, both, or neither? As we have seen, the firing rate of a typical neuron, namely, the average number of neural action potentials (or spikes) per unit time, varies (non-linearly) as a monotonic function of the intensity of a stimulus. Thus, for example, as the weight of an object you are holding increases (or decreases), the firing rate of the neurons in your biceps increases (or decreases) (cf., Adrian & Zotterman 1926). This makes rate coding an analog representation scheme. Sometimes, the precise timing of neural action potentials and, more generally, their temporal structure makes a difference to neurocognitive function. Thus, for example, as your auditory system receives sound waves from an object in the environment, the differences between the timing of the spike trains produced by distinct auditory nerve fibres in your ears enable your nervous system to identify the location of the object (cf., Konishi 2003). Temporal codes also count as a candidate analog representation scheme.

One problem with the conclusion that the brain computes with analog representations is that, traditionally, the action potential has been viewed similarly to the binary pulses of a digital computer, as a neuron either fires or not: the presence of an action potential would thus encode the binary digit 1, while its absence the digit 0 (McCulloch & Pitts 1943). But, on a closer look, neuronal spikes are not digital states. Unlike in a digital representation, where the exact relative position of each 1 and 0 within a string radically alters what is represented and what computation is performed, the precise position of an individual spike within a spike train makes very little, if any, difference. What appears to matter to neural computation is either the firing rate or the timing of spikes. While those are arguably analog representations in a broad sense, they are not continuous variables, which are the typical variables being integrated by traditional analog computers. Thus, there are disanalogies between neural computations and both digital and (paradigmatic) analog computations. Because of this, some have argued that neural computation is *sui generis* (e.g., Piccinini and Bahar 2013). At any rate, computational neuroscientists have developed techniques and formalisms that are quite different from the tools used for understanding either digital or (paradigmatic) analog computers (e.g., Dayan & Abbott 2005).

This discussion should clarify that the brain need *not* be a digital computing system performing discrete operations on digital representations in some discrete amount of time. Brains perform neural computations over neural representations. In fact, nervous systems might perform different styles of computation to carry out different functions constituting different mental capacities or, perhaps, they might approximate some aspects of a digital system at a macro-scale while the underlying computations and representations are not digital. Asking whether brains are computing systems without qualifications is less productive than asking: what sort of computing systems are brains (or parts thereof)? How should the function underlying a given mental capacity be characterised? Is it computable? If so, how could it be computed by a given neural structure?

As should be clear by now, many mental capacities can be fruitfully under-stood in neurocomputational terms, which yield mathematically precise and testable predictions and explanations. For instance, in the domain of perception, capacities like depth perception, contour integration, colour perception, motion perception, and the perception of the location of a sound have been modelled in neurocomputational terms as probabilistic, causal inferences performed by neural circuits (Pouget et al. 2013). Not all these computational models construe perception as a 'passive accumulation' and transformation of basic perceptual features; many construe perception as active, predictive, and in the service of controlling an embodied agent whose actions generate and sculpt the stream of sensory data it receives (Clark 2013). These computational models have been tested against neural and behavioural data. Their empirical success demon-strates that neural representation and computation are part of a viable empirical and theoretical scientific framework. Still, there remains the question of whether all this neural computing and representing can account for the semantic properties traditionally associated with the mind – first and foremost, intentionality.

6.4 Intentionality, Semantic Content, and Understanding

Mental states are directed towards some object or state of affairs. For example, the belief that it is raining is directed towards the putative fact that it is raining – whether or not it is actually raining. Franz Brenanto (1838–1917) re-introduced the term 'intentionality' within contemporary philosophy and used it for the directedness, or aboutness, of mental states. Brentano argued that all and only mental phenomena have intentionality in the sense that there is always an object or state of affairs that a mental state is about (1874). For example, thoughts are directed at what is thought; perception is directed at what is perceived; love is directed at what is loved. It is important not to confuse this notion of intention-ality with the more familiar notion of having an intention, purpose, or goal when one acts. The latter is just one species of the former.

Public representations such as linguistic utterances, street signs, and flags may also be seen as having intentionality because they are also about some object or state of affairs. Yet, there seems to be a big difference. Public representations appear to have *derivative* intentionality – their semantic content is there only relative to some competent observer who interprets them correctly. Utterances in Swahili are only intelligible to people who understand Swahili, while utterances in Urdu are only intelligible to people who understand Urdu, and so forth. In contrast, the intentionality of the mental state of, say, Swahili speakers who intend to utter a Swahili sentence has its intentionality (semantic content) whether or not any external observer interprets it. Thus, it seems that mental states have *original* intentionality – intentionality that is not derived from any external interpreter (Haugeland 1998, 2002; Colombo 2010).

Brentano considered (original) intentionality as a unique and irreducible property of the mind. If intentionality is irreducible in the sense that we cannot explain it in terms of non-intentional mechanisms and processes, then CTM seems insufficient to fully explain the mind. Computational mechanisms and processes would fall short of accounting for how intentionality works, why only mental states possess it, which mental states possess it, and what the structure of different intentional states is. More specifically, insofar as intentionality determines the semantic content of mental states and enables understanding, CTM seems insufficient for semantic content and understanding.

One version of the intentionality challenge against (stronger versions of) CTM is that a system running a computer program on syntactically structured vehicles will *not* thereby come to possess (original) semantic properties, in virtue of which the system's vehicles are meaningful and the system acquires the capacity for understanding; therefore, some have argued that computer symbols mean nothing on their own and computers do not understand what they are doing and do not possess intentionality. Turing (1948) formulates this sort of objection by asking us to imagine 'a paper machine for playing chess'. Variations on this thought experiment include Anatoly Mickevich's (1961) 'Game', Stanisław Lem's (1964) 'Gramophone', and John Searle's (1980) 'Chinese room'. The 'paper machine' in Turing's example consists of a list of instructions for playing chess and a human being who does not (otherwise) know how to play chess. All the human does is mindlessly follow the instructions for generating moves on the chessboard based on the current state of the board. That is, the human manipulates the symbols precisely the way a computer would. The input–output strings manipulated by the human, such as '1. d4 Nf6', '2. c4 e6', '3. Nc3 Bb4', and so forth mean nothing to the human.

Suppose you cannot directly see the human but can witness the moves on the chessboard. Would you conclude that whoever is making those moves understands how to play chess? Does the chess player perceive a knight fork as a knight fork? If you answer 'No', then intentionality and understanding are not reducible to syntactic manipulation. No matter how much the behaviour of the human resembles that of someone with genuine understanding of chess, they will never understand. If you answer 'Yes', then you might draw attention to the whole system, including the paper with the instructions written on it, devices for sensing the current state of the board and issuing commands, of which the human is part, and say that the whole system does understand how to play chess. Or you might insist that, while a mere computer running a program may not understand anything, at least some computing systems, properly connected by sensors and effectors to their environments, do possess semantic properties. Perhaps you will draw attention to the fact that cognitive scientists routinely pick out and explain the mental phenomena exhibited by computing systems by referring to their semantic properties. Finally, you might simply deny that semantic properties (including intentionality) are needed to explain mental phenomena. As we explained earlier, there are versions of CTM that reject representationalism – some argue that the semantic content ascribed by cognitive scientists in their computational explanations of mental capacities plays at best a helpful heuristic role as a 'gloss' on the mathematical, genuinely explanatory, characterisation of the target capacity, which does not involve (original) intentionality (e.g., Dennett 1978; Egan 2014, 2018).

The debate about cases like Turing's 'paper machine' has been inconclusive and often confusing, as it relies on the ill-defined notion of understanding and on an intuitive distinction between *simulating* a mental capacity and *genuinely possessing* it (Searle 1980). As we explained earlier, when we introduced the Turing test, Turing himself anticipated these difficulties. He suggested that an appropriately programmed computer or a computer with the right learning history (or 'education', as Turing put it) could perform a task like playing chess or having a conversation sufficiently well to fool us into thinking that it is a human. Turing suggested that we would take such a computer to be intelligent like us. But to many others, the question remains as to whether such a computer genuinely understands and whether its states have original intentionality.

Another approach asks what it would take for computational vehicles, including neural representations, to acquire original intentionality or at least some form of original semantic content (Morgan & Piccinini 2018; Lee 2021). This is sometimes called 'the symbol grounding problem' (Harnad 1990). In response, a number of accounts have been proposed, of which we will mention two especially influential ones. *Functional role semantics* holds that systems of vehicles that stand in the right sorts of functional relations with environmental inputs, motor outputs, and one another possess original semantic properties (e.g., Sellars 1963; Harman 1973). *Informational teleosemantics* and cognate views hold that internal vehicles that have the (teleological) function of carrying information about a certain state of affairs, and perhaps have the function of guiding behaviour on that basis, have that state of affairs as their (original) semantic content (e.g., Neander 2017; Shea 2018; Millikan 2023). These approaches might need to be augmented by an account of how signals can acquire and transmit semantic information (Skyrms 2010; Isaac 2019) and how representations and computations are acquired by fully situated (embodied, embedded, enactive, and affective) neurocognitive systems. Unlike representations in ordinary digital computers, which may or may not be situated in the right way and therefore may lack original semantic content, suitably situated neural representations might be able to acquire original semantic content (Piccinini 2022).

A final challenge to CTM related to understanding concerns common sense reasoning. It says that CTM cannot account for common sense because common sense cannot be captured with formal rules expressed in a propositional format, as it generally involves ambiguity, indeterminacy, unpredictability, and context-sensitive background knowledge. Artificial computing systems that rely solely on symbolic representations carrying propositional content, or even machine learning algorithms for prediction, pattern recognition, and clustering, could never capture this kind of common sense reasoning (Dreyfus 1992; Davis & Marcus 2015; Birhane 2021).

One way of starting to address this challenge is to notice that all representational vehicles involve compression of information, which means that some information gets lost and some information is retained in different kinds of vehicles; therefore, different representational vehicles may be more or less suitable to support different mental capacities. Language-like representations, for example, are good at conveying information about discrete objects and properties at a high level of abstraction. Iconic representations are better with concrete, perceptual information about particular objects and properties. Vectorial representations embodied in a trained neural network excel at pattern recognition and completion. Abandoning the assumption that, if brains compute, then they must rely on just one specific kind of neural code might help us better appreciate how and when our common sense reasoning relies on different types of information. In fact, some aspects of human common sense reasoning have been successfully explained in computational terms, drawing on a variety of algorithmic processes and representational vehicles.

In summary, even if CTM needs to be augmented by an account of intentionality, semantic content, and understanding, the mind can still be a computing system. While a full account of the mind might plausibly require more than computation, the challenge of intentionality does not refute CTM. A feature of the mind that strikes some as even less conducive to a computational account than intentionality is phenomenal consciousness, to which we now turn.

6.5 Consciousness

Many mental states and processes are *not* conscious, but others are conscious. This can mean a number of things. One notion of consciousness revolves around having mental states – typically, intentional mental states – that are accessible to many mental processes and can guide action, including verbal reports that express them, as when we assert what we believe. Paradigmatic examples include (conscious) beliefs and desires. Having states of this sort is sometimes called *access consciousness* (Block 1995). Another notion of consciousness revolves around having subjective experiences, also known as states with a subjective feel, phenomenal character, qualia, or what it is like to be someone. Examples include the feeling of pain or pleasure. Having states of this sort is usually called *phenomenal consciousness*. States that are phenomenally conscious are often also access conscious – for example, we can report on our pains and pleasures – and vice versa, although it is controversial whether this is always the case.

Access consciousness seems relatively approachable within CTM, especially if CTM is augmented by an account of intentionality (Section 6.4). This is because access consciousness seems to be a matter of having the right sorts of intentional states plus the right sorts of computational processes for manipulating, and possibly expressing, such intentional states. Phenomenal consciousness seems more difficult to account for.

Some argue that the phenomenal character of conscious experiences cannot be explained in purely physical, functional, or computational terms since it seems conceivable that an organism or system that is physically, functionally, or computationally equivalent to a conscious organism or system might *not* have any phenomenally conscious experiences (Leibniz 1714/1965, section 17; Block 1978; Maudlin 1989; Chalmers 1996). If it is possible, in principle, for there to be an entity that has all your physical, functional, or computational properties while lacking phenomenal experiences, then phenomenal consciousness is not a physical, functional, or computational property and we should look elsewhere for explaining it.

This argument targets not only (strong) versions of CTM, according to which *all* aspects of the mind are computational, but also any version of functionalism or physicalism. The argument is that consciousness cannot be reduced to, or explained by, any physical, functional, or computational structure. If this argument succeeds, any physicalist or functionalist account of the mind is at best incomplete since having certain physical properties, or performing certain functions, would be insufficient for possessing conscious experiences. If (physical) computation consists in performing certain functions or having certain physical properties, then CTM is insufficient to explain all aspects of mind and particularly its phenomenal character.

Even if something beyond computation is required to explain phenomenal consciousness, however, CTM could still explain mental capacities such as sensing, acting, thinking, learning, reasoning, and decision-making. Furthermore, the right computations might still be sufficient for phenomenal consciousness in a more limited sense. Consider what would happen if each one of your neurons were progressively replaced by computationally equivalent microchips (Zuboff 1981). If you guessed that you would retain your phenomenal consciousness (whether or not the computations themselves constitute the experiences), then you might be inclined to believe that some artificial computing systems, given the right structure, could be conscious. The question of what it would take for a system to be conscious remains highly controversial, particularly when it comes to artefacts (e.g., Dehaene et al. 2017).

While many are compelled by the intuition that no combination of physical, functional, or computational properties could possibly account for phenomenal consciousness, others, including many cognitive neuroscientists, have proposed accounts of phenomenal consciousness in broadly computational terms. There are now several such accounts, with different explanatory scope and targets (Seth & Bayne 2022). Ideally, a comprehensive account of consciousness would identify the neural mechanisms and processes, whether computational or not, that give rise to phenomenal consciousness. It would explain why and how some organisms and systems are phenomenally conscious but others are not – for example, why you have subjective experiences, while rocks lack them. It would also explain why and how different experiences have a different phenomenal character – for example, how and why what it is like to smell weed differs from what it is like to smell coffee, or how and why the phenomenal character of my perception that the sky is blue differs from the phenomenal character of my perception that my dog is white. Furthermore, a comprehensive account of consciousness would answer questions about the nature, phenomenological structure, and physical mechanisms of *self*-consciousness and would relate phenomena like wakefulness and behavioural responsiveness to the subjectivity and intentionality of conscious mental states, making suggestions about whether and how the phenomenal character of conscious experiences plays any causal role in the evolution and development of a subject's mental capacities. Much recent empirical work pursues these aims (e.g., Bourdillon et al. 2020; Gilson et al. 2023; Huang et al. 2023; Luppi et al. 2023).

Broadly, physicalist theories of phenomenal consciousness appeal to various sorts of physical, biophysical, neural, computational, informational, or representational properties, as well as to the situatedness (embodiment, embeddedness, and enaction) of the mind. An especially influential family of accounts that focus primarily on informational and computational properties is *global workspace theories* (Baars 1993; Mashour et al. 2020). Reminiscent of blackboard architectures in AI, where a centralised data structure is used by specialised modules to share and process information, global workspace theories claim that a subject's phenomenally conscious experience is constituted by the global availability of the subject's mental states to various cognitive processes underlying attention, memory, verbal reports, and so forth. Specifically, sensory information becomes phenomenally conscious when it is broadcast within a unified network of many modules to support verbal reports and to guide behaviour flexibly.

There are many alternatives to global workspace theories. Two alternatives that put a large emphasis on informational and representational properties are the following. *Integrated information theory* proposes that systems are phenomenally conscious to the degree that they instantiate a certain kind of causal structure (e.g., Tononi et al. 2016). The phenomenal character of a subject's experiences is identical to certain causal properties of a system that generate irreducible maxima of integrated information within the system. Integrated information is an information-theoretic quantity, which depends on the extent that a system's components are connected globally and non-redundantly. In contrast, *higher order theories* of consciousness propose that mental states become phenomenally conscious just in case they become the target of meta-representations (e.g., Lau 2022). For example, when a (first-order) perceptual state such as the sight of an avocado becomes represented as such by the system (as in 'I am seeing this avocado'), the original state – the sight of the avocado – becomes phenomenally conscious, while the meta-representation – the representation *that I am seeing this avocado* – may well remain unconscious.

Theories of phenomenal consciousness remain controversial and raise several outstanding questions (Michel et al. 2019) – for example, about how we should empirically test them and whether they make testable predictions about conscious experience in different organisms, infants, patients with brain damage, and non-biological systems. One way consciousness researchers have tried to bring precision to these and other theories leverages computational modelling, which could bridge abstract notions, concrete mechanistic features of biological brains, and organisms' behaviour, and serve as a common mathematical and theoretical language for comparing distinct theories of consciousness.

Yet, computational modelling of conscious experience also faces the challenges of how to formalise phenomenal character, how to relate it to the properties ascribed by a computational model to its target system, and how to pursue explanations – of, say, psychiatric conditions – that genuinely integrate computational and mechanistic explanations with phenomenological analyses (Colombo & Heinz 2019). For example, a general computational framework for studying consciousness is *active inference* within predictive processing accounts of the mind, whose central thesis is that a system's conscious experiences should be understood in terms of the system's ability to predict its future sensory states, including the consequences of its own actions (e.g., Friston 2018). It remains contentious whether and how this view will advance our understanding of consciousness and, more generally, how the active-inference framework should be refined and informed by data about subjective reports for productively informing consciousness research (Vilas, Auksztulewicz & Melloni 2022).

Even though there is little consensus on the nature of phenomenal consciousness, recent work on the metaphysics of mind helps clarify what the options are. On one end of the spectrum is the view that phenomenal consciousness is something non-physical. On the other end of the spectrum are the views that phenomenal consciousness either reduces to the right sort of computation and information processing or is an illusion (Dennett 1991; Frankish 2017). In between are views according to which phenomenal consciousness is an aspect of the physical qualities (i.e., intrinsic properties) of systems with the right sort of organisation, and this qualitative aspect of such systems goes beyond their purely computational character (e.g., Piccinini 2020, Chapter 14; Anderson & Piccinini forthcoming, Chapter 9).

In summary, even if CTM needs to be further augmented by an account of phenomenal consciousness, the mind can still be a computing system. While a full account of phenomenal consciousness plausibly requires more than computation and information processing, CTM remains the most powerful framework for explaining mental phenomena – or at least mental phenomena short of a complete account of their phenomenal character.

7 Conclusions

The computational theory of mind says that minds are computing systems. On its own, this claim does not rule out that minds do other things besides computing and is just the beginning of a fully fledged theory that could make empirical predictions, yield insight into the workings of the mind, and be evaluated for its degree of empirical support. While it is contentious what, exactly, a computing system is, which functions nervous systems might compute to produce specific mental phenomena, how they might do that, and whether there is a single, universally true answer to all these questions, specific versions of CTM and specific computational models of target systems and capacities have been very fruitful in pursuing many scientific goals.

References

Aaronson, D., Grupsmith, E., & Aaronson, M. (1976). The impact of computers on cognitive psychology. *Behavioral Research Methods & Instrumentation*, *8*: 129–38.

Aaronson, S. (2013). Why philosophers should care about computational complexity. In Copeland, B. J., Posy, C. J., & Shagrir, O. (eds.), *Computability: Turing, Gödel, Church, and Beyond*. Cambridge, MA: MIT Press, pp. 261–328.

Abraham, T. H. (2018). Cybernetics. In M. Sprevak & M. Colombo (eds.), *The Routledge handbook of the computational mind*. New York: Routledge, pp. 52–64.

Adamatzky, A. (2021). *Handbook of Unconventional Computing*. Singapore: World Scientific.

Adrian, E. D., & Zotterman, Y. (1926). The impulses produced by sensory nerve endings: Part 3. Impulses set up by touch and pressure. *The Journal of Physiology*, *61*(4): 465–93.

Anderson, N. G., & Piccinini, G. (forthcoming). *The Physical Signature of Computation: A Robust Mapping Account*. Oxford: Oxford University Press.

Ashby, W. R. (1952). *Design for a Brain*. London: Chapman and Hall.

Baars, B. J. (1993). *A Cognitive Theory of Consciousness*. Cambridge: Cambridge University Press.

Baker, B., Lansdell, B., & Kording, K. P. (2022). Three aspects of representation in neuroscience. *Trends in Cognitive Sciences*, *26*(11): 942–58.

Barlow, H. B. (1961). Possible Principles Underlying the Transformation of Sensory Messages. In Rosenblith, W. A. (ed.), *Sensory Communication*. Cambridge, MA: MIT Press.

Bechtel, W., & Shagrir, O. (2015). The non-redundant contributions of Marr's three levels of analysis for explaining information-processing mechanisms. *Topics in Cognitive Science*, *7*(2): 312–22.

Beer, R. D. (2000). Dynamical approaches to cognitive science. *Trends in Cognitive Sciences*, *4*(3): 91–9.

Beer, R. D., & Williams, P. L. (2015). Information processing and dynamics in minimally cognitive agents. *Cognitive Science*, *39*(1): 1–38.

Bell, A. J. (1999). Levels and loops: The future of artificial intelligence and neuroscience. *Philosophical Transactions of the Royal Society of London. Series B: Biological Sciences*, *354*(1392): 2013–20.

Bengio, Y., Lecun, Y., & Hinton, G. (2021). Deep learning for AI. *Communications of the ACM*, *64*(7): 58–65.

Bennett, M. R., & Hacker, P. M. S. (2022). *Philosophical Foundations of Neuroscience*. 2nd ed. Hoboken: John Wiley & Sons.

Bickhard, M. H., & Terveen, L. (1995). *Foundational Issues in Artificial Intelligence and Cognitive Science: Impasse and Solution*. Amsterdam: North-Holland.

Birhane, A. (2021). The impossibility of automating ambiguity. *Artificial Life*, *27*(1): 44–61.

Bishop, C. M. (2006). *Pattern Recognition and Machine Learning*. New York: Springer.

Block, N. (1978). Troubles with functionalism. In Savage, C. W. (ed.), *Perception and Cognition: Issues in the Foundations of Psychology, Minnesota Studies in the Philosophy of Science*, vol. 9. Minneapolis: University of Minnesota Press, pp. 261–325.

Block, N. (1995). On a confusion about a function of consciousness. *The Behavioral and Brain Sciences*, *18*(2): 227–87.

Block, N., & Fodor, J. A. (1972). What psychological states are not. *The Philosophical Review*, *81*(2): 159–81.

Bourdillon, P., Hermann, B., Guénot, M., et al. (2020). Brain-scale cortico-cortical functional connectivity in the delta-theta band is a robust signature of conscious states: An intracranial and scalp EEG study. *Scientific Reports*, *10*: 14037. https://doi.org/10.1038/s41598-020-70447-7.

Bowers, J. S., Malhotra, G., Dujmović, M., et al. (2022). Deep problems with neural network models of human vision. *Behavioral and Brain Sciences*, 1–74. https://doi.org/10.1017/S0140525X22002813.

Brentano, F. (1874/1973). *Psychology from an Empirical Standpoint*. Trans. Rancurello, A. C., Terrell, D. B., & McAlister, L. L. London: Routledge and Kegan Paul.

Brette, R. (2019). Is coding a relevant metaphor for the brain? *Behavioral and Brain Sciences*, *42*: e215.

Brooks, R. A. (1991). Intelligence without representation. *Artificial Intelligence*, *47*(1–3): 139–59.

Buckner, C. (forthcoming). *Deeply Rational Machines*. Oxford: Oxford University Press.

Calvo, P., & Symons, J. (eds.). (2014). *The Architecture of Cognition: Rethinking Fodor and Pylyshyn's Systematicity Challenge*. Cambridge: MIT Press.

Camp, E. (2007). Thinking with maps. *Philosophical Perspectives*, *21*: 145–82.

Campbell, D. I., & Yang, Y. (2021). Does the solar system compute the laws of motion? *Synthese*, *198*: 3203–20.

Cao, R. (2018). Computational explanations and neural coding. In Sprevak, M., & Colombo, M. (eds.), *The Routledge Handbook of the Computational Mind*. Routledge: New York, pp. 283–96.

Cao, R. (2022). Multiple realizability and the spirit of functionalism. *Synthese*, *200*: 506. https://doi.org/10.1007/s11229-022-03524-1.

Chalmers, D. J. (1994). On implementing a computation. *Minds and Machines*, 4(4): 391–402.

Chalmers, D. J. (1996). *The Conscious Mind: In Search of a Fundamental Theory*. Oxford: Oxford University Press.

Chalmers, D. J. (2011). A computational foundation for the study of cognition. *The Journal of Cognitive Science*, *12*: 323–57.

Chater, N., Tenenbaum, J. B., & Yuille, A. (2006). Probabilistic models of cognition. *Trends in Cognitive Science*, *10*(7): 287–93.

Chemero, A. (2011). *Radical Embodied Cognitive Science*. Cambridge: MIT Press.

Chirimuuta, M. (2018). Explanation in computational neuroscience: Causal and non-causal. *The British Journal for the Philosophy of Science*, *69*: 849–80.

Chirimuuta, M. (2021). Your brain is like a computer: Function, analogy, simplification. In Calzavarini, F., & Viola, M. (eds.), *Neural Mechanisms: Studies in Brain and Mind*, vol. 17. Cham: Springer, pp. 235–61.

Church, A. (1936a). A note on the Entscheidungsproblem. *Journal of Symbolic Logic*, *1*: 40–1. https://doi.org/10.2307/2269326.

Church, A. (1936b). An unsolvable problem of elementary number theory. *American Journal of Mathematics*, *58*: 345–63. https://doi.org/10.2307/2371045.

Churchland, P. M. (1992). *A Neurocomputational Perspective: The Nature of Mind and the Structure of Science*. Cambridge: MIT Press.

Churchland, P. S., & Sejnowski, T. J. (1992). *The Computational Brain*. Cambridge: MIT Press.

Cisek, P. (2019). Resynthesizing behavior through phylogenetic refinement. *Attention, Perception, & Psychophysics*, *81*(7): 2265–87.

Clark, A. (1993). *Associative Engines: Connectionism, Concepts, and Representational Change*. Cambridge: MIT Press.

Clark, A. (1997). The dynamical challenge. *Cognitive Science*, *21*(4): 461–81.

Clark, A. (1998). *Being There: Putting Brain, Body, and World Together Again*. Cambridge: MIT Press.

Clark, A. (2008). *Supersizing the Mind: Embodiment, Action, and Cognitive Extension*. New York: Oxford University Press.

Clark, A. (2013). Whatever next? Predictive brains, situated agents, and the future of cognitive science. *Behavioral and Brain Sciences*, *36*(3): 181–204.

Coelho Mollo, D. (2018). Functional individuation, mechanistic implementation: The proper way of seeing the mechanistic view of concrete computation. *Synthese*, *195*(8): 3477–97.

Collins, A. G. E., & Cockburn, J. (2020). Beyond dichotomies in reinforcement learning. *Nature Reviews Neuroscience*, *21*, 576–86.

Colombo, M. (2009). Does embeddedness tell against computationalism? A tale of bees and sea hares. *AISB09 Proceedings of the 2nd Symposium on Computing and Philosophy*. Edinburgh: Society for the Study of Artificial Intelligence and the Simulation of Behaviour, pp. 16–21.

Colombo, M. (2010). How 'authentic intentionality' can be enabled: A neurocomputational hypothesis. *Minds & Machines*, *20*: 183–202.

Colombo, M. (2014a). Deep and beautiful: The reward prediction error hypothesis of dopamine. *Studies in History and Philosophy of Science Part C: Studies in History and Philosophy of Biological and Biomedical Sciences*, *45*: 57–67.

Colombo, M. (2014b). Explaining social norm compliance: A plea for neural representations. *Phenomenology and the Cognitive Sciences*, *13*(2): 217–38.

Colombo, M. (2017). Why build a virtual brain? Large-scale neural simulations as jump start for cognitive computing. *Journal of Experimental & Theoretical Artificial Intelligence*, *29*(2): 361–70.

Colombo, M. (2021). (Mis) computation in computational psychiatry. In Calzavarini, F., & Viola, M. (eds.), *Neural Mechanisms: Studies in Brain and Mind*, vol. 17. Cham: Springer, pp. 427–48.

Colombo, M. (2022). Computational modelling for alcohol use disorder. *Erkenntnis*. https://doi.org/10.1007/s10670-022-00533-x.

Colombo, M., & Heinz, A. (2019). Explanatory integration, computational phenotypes, and dimensional psychiatry: The case of alcohol use disorder. *Theory & Psychology*, *29*(5): 697–718.

Copeland, B. J. (1996). What is computation? *Synthese*, *108*: 335–59.

Copeland, B. J. (2000). Narrow versus wide mechanism: Including a re-examination of Turing's views on the mind–machine issue. *The Journal of Philosophy*, *97*: 5–32.

Copeland, B. J., & Proudfoot, D. (1996). On Alan Turing's anticipation of connectionism. *Synthese*, *108*(3): 361–77.

Corabi, J., & Schneider, S. (2012). The metaphysics of uploading. *Journal of Consciousness Studies*, *19*(7): 26–44.

Crick, F. (1989). The recent excitement about neural networks. *Nature*, *337*: 129–32.

D'Angelo, E., & Jirsa, V. (2022). The quest for multiscale brain modeling. *Trends in Neurosciences*, *45*(10): 777–90.

Cummins, R. (1983) *The Nature of Psychological Explanation*. Cambridge: MIT Press.

Daston, L. (1994). Enlightenment calculations. *Critical Inquiry, 21*(1): 182–202.

Davis, E., & Marcus, G. (2015). Commonsense reasoning and commonsense knowledge in artificial intelligence. *Communications of the ACM, 58*(9): 92–103.

Dabney, W., Kurth-Nelson, Z., Uchida, N., et al. (2020). A distributional code for value in dopamine-based reinforcement learning. *Nature, 577*(7792): 671–75.

Daw, N. D., Niv, Y., & Dayan, P. (2005). Uncertainty-based competition between prefrontal and dorsolateral striatal systems for behavioral control. *Nature Neuroscience, 8*(12): 1704–11.

Daw, N. D. & Frank M. J. (2009). Reinforcement learning and higher level cognition: introduction to the special issue. *Cognition, 113*: 259–61.

Dayan, P. (1994). Computational modelling. *Current Opinion in Neurobiology, 4*(2): 212–17.

Dayan, P. (2001). *Levels of Analysis in Neural Modeling*. Encyclopedia of Cognitive Science. London: MacMillan Press.

Dayan, P., & Abbott, L. F. (2005). *Theoretical Neuroscience: Computational and Mathematical Modeling of Neural Systems*. Cambridge: MIT Press.

Dehaene, S., Lau, H., & Kouider, S. (2017). What is consciousness, and could machines have it? *Science, 358*(6362): 486–92.

Dennett, D. C. (1969). *Content and Consciousness*. London: Routledge & Kegan Paul.

Dennett, D. C. (1978). *Brainstorms: Philosophical Essays on Mind and Psychology*. Montgomery: Bradford.

Dennett, D. C. (1991a). Real patterns. *Journal of Philosophy, 88*(1): 27–51.

Dennett, D. C. (1991b). *Consciousness Explained*. Boston: Little, Brown.

Dewhurst, J. (2018). Individuation without representation. *The British Journal for the Philosophy of Science, 69*(1): 103–16.

Dickinson, A. (1985). Actions and habits: the development of behavioural autonomy. *Philosophical Transactions of the Royal Society of London. Series B: Biological Sciences*, 308, 67–78.

Dickinson A., & Balleine B. (2002). The role of learning in the operation of motivational systems. In: Gallistel C. R. (ed) *Stevens' handbook of experimental psychology: learning, motivation, and emotion*. New York: Wiley, pp. 497–534.

Dolan, R. J., & Dayan, P. (2013). Goals and habits in the brain. *Neuron, 80*(2): 312–25.

Dretske, F. (1981). *Knowledge and the Flow of Information*. Cambridge, MA: MIT Press.

Dreyfus, H. (1992). *What Computers Still Can't Do: A Critique of Artificial Reason*. New York: MIT Press.

Dreyfus, H. (2002a). Intelligence without representation: Merleau-Ponty's critique of mental representation. *Phenomenology and the Cognitive Sciences*, *1*(4): 413–25.

Dreyfus, H. (2002b). Refocusing the question: Can there be skillful coping without propositional representations or brain representations? *Phenomenology and the Cognitive Sciences*, *1*: 413–25.

Edelman, G. (1992). *Bright Air, Brilliant Fire*. New York: Basic Books.

Egan, F. (2014). How to think about mental content. *Philosophical Studies*, *170*(1): 115–35.

Egan, F. (2018). The nature and function of content in computational models. In Sprevak, M., & Colombo, M. (eds.),*The Routledge Handbook of the Computational Mind*. New York: Routledge, pp. 247–58.

Eliasmith, C. (1996). The third contender: A critical examination of the dynamicist theory of cognition. *Philosophical Psychology*, *9*(4): 441–63.

Eliasmith, C. (2009). How we ought to understand computation in the brain. *Studies in History and Philosophy of Science*, *41*: 313–20.

Eliasmith, C., Stewart, T. C., Choo, X., et al. (2012). A large-scale model of the functioning brain. *Science*, *338*(6111): 1202–5.

Evans, J. S. B., & Stanovich, K. E. (2013). Dual-process theories of higher cognition advancing the debate. *Perspectives on Psychological Science*, *8*(3): 223–41.

Faisal, A. A., Selen, L. P., & Wolpert, D. M. (2008). Noise in the nervous system. *Nature Reviews Neuroscience*, *9*(4): 292–303.

Feynman, R., Leighton R. B., & Sands, M. L. (1989). *Lectures on Physics*, vol. 1, retrieved from Caltech, HTML Edition. www.FeynmanLectures.caltech.edu/.

Figdor, C. (2018). *Pieces of Mind: The Proper Domain of Psychological Predicates*. New York: Oxford University Press.

Fodor, J. A. (1965). Explanations in psychology. In Black, M. (ed.), *Philosophy in America*. London: Routledge & Kegan Paul, pp. 161–79.

Fodor, J. A. (1968). *Psychological Explanation*. New York: Random House.

Fodor, J. A. (1975). *The Language of Thought*. New York: Thomas Y. Crowell.

Fodor, J. A. (1987). *Psychosemantics*. Cambridge: MIT Press.

Fodor, J. A. (1996). Deconstructing Dennett's Darwin, *Mind and Language*, *11*: 246–62.

Fodor, J. A., & Pylyshyn, Z. W. (1988). Connectionism and cognitive architecture: A critical analysis. *Cognition*, *28*(1–2): 3–71.

Frankish, K. (ed.). (2017). *Illusionism: As a Theory of Consciousness*. Exeter: Imprint Academic.

Freeman, W. J. (1991). The physiology of perception. *Scientific American, 264*: 78–85.

Fresco, N. (2014). *Physical Computation and Cognitive Science*. Heidelberg: Springer.

Fresco, N. (2022). Information in explaining cognition: How to evaluate it? *Philosophies, 7*(2): 28.

Fresco, N., Copeland, B. J., & Wolf, M. J. (2021). The indeterminacy of computation. *Synthese, 199*(5): 12753–75.

Fresco, N., & Primiero, G. (2013). Miscomputation. *Philosophy & Technology, 26*(3): 253–72.

Friston, K. (2018). Am I self-conscious?(Or does self-organization entail self-consciousness?). *Frontiers in Psychology, 9*: 579.

Gabor, D. (1946). Theory of communication. Part 1: The analysis of information. *Journal of the Institution of Electrical Engineers-part III: Radio and Communication Engineering, 93*(26): 429–41.

Gallagher, S. (2006). *How the Body Shapes the Mind*. New York: Oxford University Press.

Gallistel, C. R. (1990). *The Organization of Learning*. Cambridge: The MIT Press.

Gallistel, C., & King, A. (2010). *Memory and the Computational Brain*. Oxford: Wiley-Blackwell.

Garey, M. R., & Johnson, D. S. (1979). *Computers and Intractability: A Guide to the Theory of NP-completeness*. San Francisco, CA: W. H. Freeman.

Gerard, R. W. (1951). Some of the problems concerning digital notions in the central nervous system: Cybernetics. In Foerster, H. V., Mead, M., & Teuber, H. L. (eds.), *Circular Causal and Feedback Mechanisms in Biological and Social Systems*. Transactions of the Seventh Conference. New York: Macy Foundation, pp. 11–57.

Gibson, J. J. (1979). *The Ecological Approach to Visual Perception*. Boston: Houghton Mifflin.

Gigerenzer, G., & Goldstein, D. G. (1996a). Mind as computer: Birth of a metaphor. *Creativity Research Journal, 9*(2–3): 131–44.

Gigerenzer, G., & Goldstein, D. G. (1996b). Reasoning the fast and frugal way: Models of bounded rationality. *Psychological Review, 103*(4): 650–69. https://doi.org/10.1037/0033-295X.103.4.650.

Gilson, M., Tagliazucchi, E., & Cofré, R. (2023). Entropy production of multivariate Ornstein-Uhlenbeck processes correlates with consciousness levels in the human brain. *Physical Review, E107*: 024121.

Gillett, C. (2007). A mechanist manifesto for the philosophy of mind: A third way for functionalists. *Journal of Philosophical Research*, *32*: 21–42.

Gładziejewski, P., & Miłkowski, M. (2017). Structural representations: Causally relevant and different from detectors. *Biology and Philosophy*, *32*: 337–55.

Globus, G. G. (1992). Toward a noncomputational cognitive neuroscience. *Journal of Cognitive Neuroscience*, *4*(4): 299–300.

Gödel, K. (1931). Über formal unentscheidbare Sätze der Principia Mathematica und verwandter Systeme, I, *Monatshefte für Mathematik und Physik*, *38*: 173–98. Reprinted in Feferman, S., Kleene, S., Moore, G., Solovay, R., & van Heijenoort, J. (eds.). (1986). *Collected Works. I: Publications 1929–1936*. Oxford: Oxford University Press, pp. 144–95.

François-Lavet, V., Henderson, P., Islam, R., Bellemare, M. G., & Pineau, J. (2018). An Introduction to Deep Reinforcement Learning. *Foundations and Trends in Machine Learning*, *11*(3–4): 219–354.

Gödel, K. (1951). Some basic theorems on the foundations of mathematics and their implications, lecture manuscript. Feferman, S., Dawson, J., Kleene, S., et al. (eds.). (1995). *Collected Works. III: Unpublished Essays and Lectures*. Oxford: Oxford University Press, pp. 304–23.

Godfrey-Smith, P. (2009). Triviality arguments against functionalism. *Philosophical Studies*, *145*: 273–95.

Goodfellow, I., Bengio, Y., & Courville, A. (2016). *Deep Learning*. Cambridge: MIT Press.

Graves, A., Wayne, G., & Danihelka, I. (2014). *Neural Turing Machines*. arXiv preprint arXiv: 1410.5401.

Griffiths, T. L., Chater, N., Kemp, C., Perfors, A., & Tenenbaum, J. B. (2010). Probabilistic models of cognition: Exploring representations and inductive biases. *Trends in Cognitive Sciences*, *14*(8): 357–64.

Guest, O., & Martin, A. E. (2021). How computational modeling can force theory building in psychological science. *Perspectives on Psychological Science*, *16*(4): 789–802.

Harman, G. (1973). *Thought*. Princeton: Princeton University Press.

Harnad, S. (1990). The symbol grounding problem. *Physica D: Nonlinear Phenomena*, *42*(1-3): 335–46.

Hassabis, D., Kumaran, D., Summerfield, C., & Botvinick, M. (2017). Neuroscience-inspired artificial intelligence. *Neuron*, *95*(2): 245–58.

Haugeland, J. (1985). *Artificial Intelligence: The Very Idea*. Cambridge, MA: MIT Press.

Haugeland, J. (1998). *Having Thought: Essays in the Metaphysics of Mind*. Cambridge: Harvard University Press.

Haugeland, J. (2002). Authentic intentionality. In Scheutz, M. (ed.), *Computationalism: New Directions*. Cambridge, MA: MIT Press, pp. 159–74.

Hebb, D. (1949). *The Organization of Behavior*. New York: Wiley & Sons.

Hesse, M. (1966). *Models and Analogies in Science*. Notre Dame: University of Notre Dame Press.

Hodgkin, A. L., & Huxley, A. F. (1952). A quantitative description of membrane current and its application to conduction and excitation in nerve. *The Journal of Physiology, 117*(4): 500–44.

Hopfield, J. J. (1982). Neural networks and physical systems with emergent collective computational abilities. *Proceedings of the National Academy of Sciences, 79*: 2554–8.

Horsman, D., Kendon, V., & Stepney, S. (2018). Abstraction/representation theory and the natural science of computation. In Cuffaro, M. E. & Fletcher S. C. (eds.), *Physical Perspectives on Computation, Computational Perspectives on Physics*. Cambridge: Cambridge University Press, pp. 127–52.

Houk, J. C., Adams, J. L., & Barto, A. G. (1995). A Model of How the Basal Ganglia Generate and Use Neural Signals that Predict Reinforcement. In J. C. Houk, J. L. Davis, D. G. Beiser (eds.), *Models of Information Processing in the Basal Ganglia*. Cambridge: MIT Press, pp. 249–70.

Huang, Z., Mashour, G. A., & Hudetz, A. G. (2023). Functional geometry of the cortex encodes dimensions of consciousness. *Nature Communications, 14*, 72. https://doi.org/10.1038/s41467-022-35764-7.

Huys, Q. J., Maia, T.V., & Frank, M.J. (2016). Computational psychiatry as a bridge from neuroscience to clinical applications. *Nature Neuroscience, 19*(3), 404–13.

Isaac, A. M. C. (2017). The semantics latent in shannon information. *The British Journal for the Philosophy of Science, 70*(1): 103–25.

Isaac, A. M. C. (2018a). Computational thought from Descartes to Lovelace. In Sprevak, M., & Colombo, M. (eds.), *The Routledge Handbook of the Computational Mind*. New York: Routledge, pp. 9–22.

Isaac, A. M. C. (2018b). Embodied cognition as analog computation. *Reti, Saperi, Linguaggi: Italian Journal of Cognitive Sciences, 2*: 239–62.

Isaac, A. M. C. 2019. The semantics latent in Shannon information. *The British Journal for the Philosophy of Science, 70*: 103–25.

Izhikevich, E. (2007). *Dynamical Systems in Neuroscience: The Geometry of Excitability and Bursting*. Cambridge, MA: The MIT Press.

Jeffress, L. A. (ed.). (1951). *Cerebral Mechanisms in Behavior*. New York: Wiley.

Jonas, E., & Körding, K. P. (2017). Could a neuroscientist understand a microprocessor? *PLoS Computational Biology*, *13*(1): e1005268.

Kaelbling, L. P., Littman, M. L., & Moore, A. W. (1996). Reinforcement learning: A survey. *Journal of artificial intelligence research*, *4*, 237–85.

Kahneman, D. (2011). *Thinking, fast and slow*. Macmillan

Kaplan, D. M. (2011). Explanation and description in computational neuroscience. *Synthese*, *183*(3): 339–73.

Kiela, D., Bartolo, M., Nie, Y., et al. (2021). *Dynabench: Rethinking Benchmarking in NLP*. arXiv preprint arXiv: 2104.14337.

Kirkpatrick, K. L. (2022). Biological computation: Hearts and flytraps. *Journal of Biological Physics*, *48*(1): 55–78.

Kleene, S. C. (1936). General recursive functions of natural numbers. *Mathematische Annelen*, *112*: 727–42.

Kleene, S. C. (1956). Representation of events in nerve nets and finite automata. In Shannon, C. E., & McCarthy, J. (eds.), *Automata Studies*. Princeton, NJ: Princeton University Press, pp. 3–42.

Klein, C. (2008). Dispositional implementation solves the superfluous structure problem. *Synthese*, *165*: 141–53.

Knill, D. C., & Pouget, A. (2004). The Bayesian brain: The role of uncertainty in neural coding and computation. *TRENDS in Neurosciences*, *27*(12): 712–19.

Koch, C., & Segev, I. (eds.). (1998). *Methods in Neuronal Modeling: From Synapses to Networks*. Cambridge: MIT Press.

Koch, C. (1999). *Biophysics of computation: information processing in single neurons*. New York: Oxford University Press.

Konishi, M. (2003). Coding of auditory space. *Annual Review of Neuroscience*, *26*: 31–55.

Körding, K. P., Blohm, G., Schrater, P., & Kay, K. (2018). *Appreciating Diversity of Goals in Computational Neuroscience*. https://doi.org/10.31219/osf.io/3vy69.

Lake, B. M., Ullman, T. D., Tenenbaum, J. B., & Gershman, S. J. (2016). Building machines that learn and think like people. *Behavioral and Brain Sciences*, *40*: 1–101. https://doi.org/10.1017/S0140525X16001837.

Langdon, A. J., Sharpe, M. J., Schoenbaum, G., & Niv, Y. (2018). Model-based predictions for dopamine. *Current Opinion in Neurobiology*, *49*: 1–7.

Landgrebe, J., & Smith, B. (2022). *Why machines will never rule the world: artificial intelligence without fear*. New York: Taylor & Francis.

Lashley, K. S. (1958). Cerebral organization and behavior. *Research Publications, Association for Research in Nervous and Mental Diseases*, *36*: 1–18.

Lau, H. (2022). *In Consciousness We Trust: The Cognitive Neuroscience of Subjective Experience*. Oxford: Oxford University Press.

Laughlin, S. B., de Ruyter van Steveninck, R. R., & Anderson, J. C. (1998). The metabolic cost of neural information. *Nature neuroscience*, *1*(1): 36–41.

LeCun, Y., Bengio, Y., & Hinton, G. (2015). Deep learning. *Nature*, *521*(7553): 436–44.

Lee, J. (2018). Structural representation and the two problems of content. *Mind and Language*, *34*: 606–26.

Lee, J. (2021). Rise of the swamp creatures: Reflections on a mechanistic approach to content. *Philosophical Psychology*, *34*: 805–28.

Leibniz, G. W. (1714). *The Monadology. Monadology and Other Philosophical Essays* (1965) translated and edited by Schrecker, P., & Schrecker, A. M. New York: Bobbs-Merrill.

Lem, S. (1964). *Summa Technologiae*. Electronic Mediations Series. Trans. Zylinska, J. (2013). Minneapolis, MN: University of Minnesota Press.

Lieder, F., & Griffiths, T. L. (2020). Resource-rational analysis: Understanding human cognition as the optimal use of limited computational resources. *Behavioral and Brain Sciences*, *43*: 1–60.

Light, J. S. (1999). When computers were women. *Technology and Culture*, *40*: 455–83.

Lillicrap, T. P., Santoro, A., Marris, L., Akerman, C. J., & Hinton, G. (2020). Backpropagation and the brain. *Nature Reviews Neuroscience*, *21*(6): 335–46.

Lovelace, A. A. (1843). Translation of, and notes to, Luigi F. Menabrea's sketch of the analytical engine invented by Charles Babbage. *Scientific Memoirs*, *3*: 691–731.

Lucas, J. R. (1961). Minds, Machines, and Gödel. *Philosophy*, *36*: 112–37.

Luppi, A. I., Vohryzek, J., Kringelbach, M. L., et al. (2023). Distributed harmonic patterns of structure-function dependence orchestrate human consciousness. *Communications Biology*, *6*: 117. https://doi.org/10.1038/s42003-023-04474-1.

Lyon, P., Keijzer, F., Arendt, D., & Levin, M. (2021). Reframing cognition: Getting down to biological basics. *Philosophical Transactions of the Royal Society B*, *376*(1820): 20190750.

MacKay, D. M. (1969). *Information, mechanism and meaning*. Cambridge: MIT Press.

Maia, T. V., & Frank, M. J. (2011). From reinforcement learning models to psychiatric and neurological disorders. *Nature Neuroscience*, *14*(2): 154–62.

Maley, C. J. (2023). Analogue computation and representation. *The British Journal for the Philosophy of Science*, 271–7. https://doi.org/10.1086/715031.

Maley, C. J., & Piccinini, G. (2016). Closed loops and computation in neuroscience: What it means and why it matters. In El Hady, A. (ed.), *Closed Loop Neuroscience*. London: Elsevier, pp. 271–7.

Maley, C. J., & Piccinini, G. (2017). A unified mechanistic account of teleological functions for psychology and neuroscience. In Kaplan, D. M. (ed.), *Explanation and Integration in Mind and Brain Science*. Oxford: Oxford University Press, 236–56.

Mandelbaum, E. (2022). Everything and more: The prospects of whole brain emulation. *The Journal of Philosophy, 119*(8): 444–59.

Marcus, G. (2018). *Deep Learning: A Critical Appraisal*. arXiv preprint arXiv: 1801.00631.

Marcus, G. F. (2001). *The Algebraic Mind: Integrating Connectionism and Cognitive Science*. Cambridge, MA: MIT Press.

Marcus, G., & Davis, E. (2019). *Rebooting AI: Building Artificial Intelligence We Can Trust*. New York: Vintage.

Marder, E., & Goaillard, J. M. (2006). Variability, compensation and homeostasis in neuron and network function. *Nature Reviews Neuroscience, 7*(7): 563–74.

Markram, H. (2006). The blue brain project. *Nature Reviews Neuroscience, 7*(2): 153–60.

Marr, D., & Poggio, T. (1976). From understanding computation to understanding neural circuitry [AI Memo 357]. *MIT Artificial Intelligence Laboratory.* https://dspace.mit.edu/bitstream/handle/1721.1/5782/AIM-357.pdf.

Marr, D. (1982) *Vision*. San Francisco: W.H. Freeman.

Mashour, G. A., Roelfsema, P., Changeux, J. P., & Dehaene, S. (2020). Conscious processing and the global neuronal workspace hypothesis. *Neuron, 105*(5): 776–98.

Maudlin, T. (1989). Computation and consciousness. *The Journal of Philosophy, 86*(8): 407–32.

McCarthy, J. (1959). Programs with common sense. In *Proceedings of the Teddington Conference on the Mechanization of Thought Processes*. London: Her Majesty's Stationary Office, pp. 75–91.

McClelland, J. L. (2009). The place of modeling in cognitive science. *Topics in Cognitive Science, 1*(1): 11–38.

McClelland, J. L., Botvinick, M. M., Noelle, D. C., et al. (2010). Letting structure emerge: Connectionist and dynamical systems approaches to cognition. *Trends in cognitive sciences, 14*(8): 348–56.

McCulloch, W. S. (1949). The brain computing machine. *Electrical Engineering, 68*(6): 492–97.

McCulloch, W. S., & Pitts W. H. (1943). A logical calculus of the ideas immanent in nervous activity. *Bulletin of Mathematical Biophysics, 7*: 115–33.

Michel, M., Beck, D., Block, N., et al. (2019). Opportunities and challenges for a maturing science of consciousness. *Nature Human Behaviour*, *3*(2): 104–7.

Mickevich, A. (1961). *The Game*. Translation of *Dneprov* (1961). Moscow: Moscow State University.

Miłkowski, M. (2013). *Explaining the Computational Mind*. Cambridge, MA: MIT Press.

Miłkowski, M. (2017). Situatedness and embodiment of computational systems. *Entropy*, *19*(4): 1–15. https://doi.org/10.3390/e19040162.

Miłkowski, M. (2018). From computer metaphor to computational modeling: The evolution of computationalism. *Minds and Machines*, *28*(3): 515–41.

Miller, G. (2003). The cognitive revolution: A historical perspective. *Trends in Cognitive Sciences*, *7*(3): 141–44.

Millikan, R. G. (2023). Teleosemantics and the frogs. *Mind & Language*, 1–9. https://doi.org/10.1111/mila.12456.

Minsky, M., & Seymour, P. (1969). *Perceptrons*. Cambridge: MIT Press.

Mnih, V., Kavukcuoglu, K., Silver, D., et al. (2015). Human-level control through deep reinforcement learning. *Nature*, 518(7540): 529.

Montague, P. R., Dayan, P, Person, C, & Sejnowski, T. J. (1995). Bee foraging in uncertain environments using predictive Hebbian learning. *Nature*, 377: 725–8.

Montague, P. R., Dolan, R. J., Friston, K. J., & Dayan, P. (2012). Computational psychiatry. *Trends in Cognitive Sciences*, *16*(1): 72–80.

Montague, R. (2007). *Your Brain Is (Almost) Perfect: How We Make Decisions*. London: Penguin.

Morgan, A. (2022). Against neuroclassicism: On the perils of armchair neuroscience. *Mind & Language*, *37*(3): 329–55.

Morgan, A., & Piccinini, G. (2018). Towards a cognitive neuroscience of intentionality. *Minds and Machines*, *28*: 119–39.

Morillo, C. (1992). Reward event systems: Reconceptualizing the explanatory roles of motivation, desire and pleasure. *Philosophical Psychology*, *5*: 7–32.

Moutoussis, M., Shahar, N., Hauser, T. U., & Dolan, R. J. (2019). Computation in psychotherapy, or how computational psychiatry can aid learning-based psychological therapies. *Computational Psychiatry*, *2*: 50–73.

Murphy, R. R. (2019). *Introduction to AI Robotics*. Cambridge: MIT Press.

Neander, K. (1991). Functions as selected effects: The conceptual analyst's defense. *Philosophy of Science*, *58*(2): 168–84.

Neander, K. (2017). *A Mark of the Mental: In Defense of Informational Teleosemantics*. Cambridge, MA: MIT Press.

Newell, A. (1982). The knowledge level. *Artificial intelligence*, *18*(1): 87–127.

Newell, A. (1990). *Unified Theories of Cognition*. Cambridge, MA: Harvard University Press.

Newell, A., & Simon, H. A. (1976). Computer science as empirical inquiry: Symbols and search. *Communications of the Association for Computing Machinery, 19*: 113–26.

Niv, Y. (2009). Reinforcement learning in the brain. *Journal of Mathematical Psychology, 53*(3): 139–54.

Niv, Y., & Montague, P. R. (2009) Theoretical and empirical studies of learning. In Glimcher, P. W., et al. (eds.), *Neuroeconomics: Decision Making and the Brain*. New York: Academic Press, pp. 249–329.

O'Neil, C. (2016). *Weapons of Math Destruction: How Big Data Increases Inequality and Threatens Democracy*. London: Broadway Books.

Papayannopoulos, P., Fresco, N., & Shagrir, O. (2022). On two different kinds of computational indeterminacy. *The Monist, 105*(2): 229–46.

Patzelt, E. H., Hartley, C. A., & Gershman, S. J. (2018). Computational phenotyping: using models to understand individual differences in personality, development, and mental illness. *Personality Neuroscience, 1*: e18.

Pavlick, E. (2022). Semantic Structure in Deep Learning. *Annual Review of Linguistics, 8*(1): 447–71.

Pearl, J. (2000). *Causality: Models, Reasoning, and Inference*. Cambridge: Cambridge University Press.

Penrose, R. (1989). *The Emperor's New Mind: Concerning Computers, Minds, and the Laws of Physics*. Oxford: Oxford University Press.

Piccinini, G. (2004). The First computational theory of mind and brain: A close look at McCulloch and Pitts's 'logical calculus of ideas immanent in nervous activity'. *Synthese, 141*: 175–215.

Piccinini, G. (2010). The mind as neural software? Understanding functionalism, computationalism, and computational functionalism. *Philosophy and Phenomenological Research, 81*(2): 269–311.

Piccinini, G. (2015). *Physical Computation: A Mechanistic Account*. Oxford: Oxford University Press.

Piccinini, G., & Bahar, S. (2013). Neural Computation and the Computational Theory of Cognition. *Cognitive Science, 34*: 453–88.

Piccinini, G. (2020). *Neurocognitive Mechanisms: Explaining Biological Cognition*. Oxford: Oxford University Press.

Piccinini, G. (2021). The myth of mind uploading. In Clowes, R. W., Gärtner, K., & Hipólito, I. (eds.), *The Mind-Technology Problem*. Studies in Brain and Mind, vol. 18. Cham: Springer.

Piccinini, G. (2022). Situated neural representations: Solving the problems of content. *Frontiers in Neurorobotics, 16*: 1–13.

Piccinini, G., and B. J. Ritchie (forthcoming). Cognitive Computational Neuroscience. In Heinzelmann, N. (ed.), *Advances in Neurophilosophy*. Bloomsbury.

Poldrack, R. A. (2021). The physics of representation. *Synthese, 199*(1): 1307–25.

Polger, T. W., & Shapiro, L. A. (2016). *The Multiple Realization Book*. New York: Oxford University Press.

Potochnik, A. (2017). *Idealization and the Aims of Science*. Chicago: University of Chicago Press.

Pouget, A., Beck, J. M., Ma, W. J., & Latham, P. E. (2013). Probabilistic brains: Knowns and unknowns. *Nature Neuroscience, 16*(9): 1170–8.

Price, C. J., & Friston, K. J. (2002). Degeneracy and cognitive anatomy. *Trends in Cognitive Sciences, 6*(10): 416–21.

Psillos, S. (2011). Living with the abstract: Realism and models. *Synthese, 180*(1): 3–17.

Putnam, H. (1960). Minds and machines. In Hook, S. (ed.), *Dimensions of Mind*. New York: New York University Press, pp. 57–80.

Putnam, H. (1967). Psychological predicates. In Capitan, W. H., & Merrill, D. D. (eds.), *Art, Philosophy, and Religion*. Pittsburgh: University of Pittsburgh Press. Reprinted as The nature of mental states. In Lycan, W. (ed.). (1999). *Mind and Cognition: An Anthology*. 2nd ed. Malden: Blackwell, pp. 27–34.

Putnam, H. (1975). The mental life of some machines. In *Mind, Language and Reality: Philosophical Papers*, vol. 2. Cambridge: Cambridge University Press, pp. 408–28.

Putnam, H. (1988). *Representation and Reality*. Cambridge, MA: MIT Press.

Pylyshyn, Z. W. (1980). Computation and cognition: Issues in the foundations of cognitive science. *Behavioral and Brain Sciences, 3*(1): *111–32*.

Pylyshyn, Z. W. (1984). *Computation and Cognition*. Cambridge: MIT Press.

Quilty-Dunn, J., Porot, N., & Mandelbaum, E. (2022). The best game in town: The re-emergence of the language of thought hypothesis across the cognitive sciences. *Behavioral and Brain Sciences*, 1–55. https://doi.org/10.1017/S0140525X22002849.

Quiroga, R. Q., & Panzeri, S. (eds.). (2013). *Principles of Neural Coding*. Boca Raton: CRC Press.

Rahwan, I., Cebrian, M., Obradovich, N., et al. (2019). Machine behaviour. *Nature, 568*(7753): 477–86.

Ramsey, W. M. (2016). Untangling two questions about mental representation. *New Ideas in Psychology, 40*: 3–12.

Rangel, A., Camerer, C., & Montague, P. R. (2008). A framework for studying the neurobiology of value-based decision making. *Nature Reviews Neuroscience, 9*(7): 545–56.

Rashevsky, N. (1938). *Mathematical Biophysics: Physicomathematical Foundations of Biology.* Chicago: University of Chicago Press.

Rescorla, M. (2014). A theory of computational implementation. *Synthese, 191*(6): 1277–307.

Richards, B. A., Lillicrap, T. P., Beaudoin, P., et al. (2019). A deep learning framework for neuroscience. *Nature Neuroscience, 22*(11): 1761–70.

Rosenblatt, F. (1958). The perceptron: A probabilistic model for information storage and organization in the brain. *Psychological Review, 65*(6): 386–408.

Rumelhart, D., McClelland, J., & the PDP Research Group. (1986). *Parallel Distributed Processing,* vol. 1. Cambridge: MIT Press.

Russell, S. J. (1997). Rationality and intelligence. *Artificial intelligence, 94*(1–2): 57–77.

Russell, S. J., & Norvig, P. (2009). *Artificial Intelligence: A Modern Approach.* Upper Saddle River: Prentice Hall.

Santoro, A., Lampinen, A., Mathewson, K., Lillicrap, T., & Raposo, D. (2021). *Symbolic Behaviour in Artificial Intelligence.* arXiv preprint arXiv: 2102.03406.

Schneider, S. (2011). *The Language of Thought: A New Philosophical Direction.* Cambridge: MIT Press.

Schroeder, T. (2004). *Three Faces of Desire,* New York: Oxford University Press.

Schultz, W., Dayan, P., & Montague, P.R. (1997). A neural substrate of prediction and reward. *Science, 275*: 1593–9.

Schwartz, E. L. (ed.). (1990). *Computational Neuroscience.* Cambridge: MIT Press.

Schweizer, P. (2019). Computation in physical systems: A normative mapping account. In Berkich, D., d'Alfonso, M. (eds.), *On the Cognitive, Ethical, and Scientific Dimensions of Artificial Intelligence.* Cham: Springer, pp. 27–47. https://doi.org/10.1007/978-3-030-01800-9_2.

Searle, J. R. (1980). Minds, brains and programs. *Behavioral and Brain Sciences, 3*(3): 417–57.

Segundo Ortín, M., & Calvo, P. (2022). Consciousness and cognition in plants. *Wiley Interdisciplinary Reviews: Cognitive Science, 13*(2): e1578.

Sejnowski, T. J., Koch, C., & Churchland, P. S. (1988). Computational neuroscience. *Science, 241*(4871): 1299–306.

Sellars, W. (1963). *Science, Perception, and Reality.* Atascadero, CA: Ridgeview.

Seth, A. K., & Bayne, T. (2022). Theories of consciousness. *Nature Reviews Neuroscience, 23*: 439–52.

Shagrir, O. (2022). *The Nature of Physical Computation.* New York: Oxford University Press.

Shah, A. (2012). Psychological and neuroscientific connections with reinforcement learning. In Wiering, M., & Otterlo, M., (eds). *Reinforcement Learning*. Berlin: Springer, pp. 507–37.

Shannon, C. (1948). A mathematical theory of communication. *Bell Systems Technical Journal, 27*: 279–423, 623–56.

Shapiro, L. (2010). *Embodied Cognition*. New York: Routledge.

Shea, N. (2018). *Representation in Cognitive Science*. Oxford: Oxford University Press.

Silberstein, M., & Chemero, A. (2012). Complexity and extended phenomenological-cognitive systems. *Topics in Cognitive Science, 4*(1): 35–50.

Simon, H. A. (1969). *Sciences of the Artificial*. Cambridge: MIT Press.

Simon, H. A. (1979). Information processing models of cognition. *Annual Review Psychology, 30*: 363–96.

Skyrms, B. (2010). *Signals: Evolution, Learning, & Information*. New York: Oxford University Press.

Smith, L. B., & Thelen, E. (2003). Development as a dynamic system. *Trends in Cognitive Sciences, 7*(8): 343–48.

Spirtes, P., Glymour, C. N., Scheines, R., & Heckerman, D. (2000). *Causation, Prediction, and Search*. Cambridge: MIT Press.

Spivey, M. (2007). *The Continuity of Mind*. Oxford: Oxford University Press.

Sporns, O. (2016). *Networks of the Brain*. Cambridge: MIT Press.

Sprevak, M. (2010). Computation, individuation, and the received view on representation. *Studies in History and Philosophy of Science Part A, 41*(3): 260–70.

Sprevak, M. (2018). Triviality arguments about computational implementation. In Sprevak, M., & Colombo, M. (eds.), *The Routledge Handbook of the Computational Mind*. Routledge: New York, pp. 175-91.

Sprevak, M., & Colombo, M. (eds.). (2018). *The Routledge Handbook of the Computational Mind*. Routledge: New York.

Sterling, P., & Laughlin, S. (2015). *Principles of Neural Design*. Cambridge: MIT Press.

Stich, S. (1983). *From Folk Psychology to Cognitive Science: The Case Against Belief*. Cambridge, MA: MIT Press.

Sutton, R. S., & Barto, A. G. (2018). *Reinforcement Learning. An Introduction*. 2nd ed. Cambridge: MIT Press.

Tenenbaum, J. B., Kemp, C., Griffiths, T. L., & Goodman, N. D. (2011). How to grow a mind: Statistics, structure, and abstraction. *Science, 331*(6022): 1279–85.

Thorndike, E. L. (1932). *The fundamentals of learning*. New York: Teachers College Press.

Tononi, G., Boly, M., Massimini, M., & Koch, C. (2016). Integrated information theory: From consciousness to its physical substrate. *Nature Reviews Neuroscience, 17*(7): 450–61.

Tucker, C. (2018). How to explain miscomputation. *Philosophers' Imprint, 18*(24): 1–17.

Turing, A. (1936). On computable numbers, with an application to the Entscheidungsproblem. *Proceedings of the London Mathematical Society, 42*: 230–65.

Turing, A. (1948). *Intelligent Machinery: A Report*. London: National Physical Laboratory.

Turing, A. (1950). Computing Machinery and Intelligence. *Mind, 49*: 433–60.

Uckelman, S. (2018). Computation in mediaeval Western Europe. In Hansson, S. O. (ed.), *Technology and Mathematics: Philosophical and Historical Investigations*. Berlin: Springer, pp. 33–46.

van Gelder, T. (1995). What might cognition be, if not computation? *The Journal of Philosophy, 92*(7): 345–81.

van Rooij, I. (2008). The tractable cognition thesis. *Cognitive Science, 32*(6): 939–84.

van Rooij, I., Wright, C. D., & Wareham, T. (2012). Intractability and the use of heuristics in psychological explanations. *Synthese, 187*(2): 471–87.

Varela, F. J., Thompson, E., & Rosch, E. (2016). *The Embodied Mind, Revised Edition: Cognitive Science and Human Experience*. Cambridge: MIT Press.

Vendler, Z. (1972). *Res cogitans*. Ithaca: Cornell University Press.

Vilas, M. G., Auksztulewicz, R., & Melloni, L. (2022). Active inference as a computational framework for consciousness. *Review of Philosophy and Psychology, 13*(4): 859–78. https://doi.org/10.1007/s13164-021-00579-w.

Villalobos, M., & Dewhurst, J. (2017). Why post-cognitivism does not (necessarily) entail anti-computationalism. *Adaptive Behavior, 25*(3): 117–28.

Villalobos, M., & Dewhurst, J. (2018). Enactive autonomy in computational systems. *Synthese, 195*(5): 1891–908.

von Neumann, J. (1951). The general and logical theory of automata. In Jeffress, L. A. (ed.), *Cerebral Mechanisms in Behavior: The Hixon Symposium*. New York: John Wiley & Sons, pp. 1–31.

von Neumann, J. (1958). *The Computer and the Brain*. New Haven: Yale University Press.

von Neumann, J. (1966). *Theory of Self-Reproducing Automata*, Burks, A. W. (ed.), Urbana: University of Illinois Press.

von Neumann, J. (1981). First draft report on the EDVAC. Report prepared for the U.S. Army Ordnance Department under contract W-670-ORD-4926, 1945. In Stern, N. (ed.), *From ENIAC to UNIVAC*. Bedford: Digital Press, pp. 177–246.

Weinberger, N., & Allen, C. (2022). Static-dynamic hybridity in dynamical models of cognition. *Philosophy of Science, 89*: 283–301.

Weisberg, M. (2013). *Simulation and Similarity: Using Models to Understand the World*. Oxford: Oxford University Press.

Weiskopf, D. A. (2018). The explanatory autonomy of cognitive models. In Kaplan, D. M. (ed.), *Explanation and Integration in Mind and Brain Science*. Oxford: Oxford University Press, pp. 44–69.

Werbos. P. (1974). *Beyond regression: New tools for prediction and analysis in the behavioral sciences*. Ph.D. thesis, Committee on Applied Mathematics. Cambridge: Harvard University.

Wiener, N. (1948). *Cybernetics*. New York: John Wiley.

Wiese, W. (forthcoming). Could large language models be conscious? A perspective from the free energy principle. In Hipolito, I., Hesp, C., & Friston, K. (eds.), *The Free Energy Principle: Science, Technology, and Philosophy*. London: Routledge.

Wilson, M. (2006). *Wandering Significance: An Essay on Conceptual Behaviour*. Oxford: Oxford University Press.

Wilson, M. (2022). *Imitation of Rigor*. Oxford: Oxford University Press.

Wilson, R. A. (1994). Wide computationalism. *Mind, 103*(411): 351–72.

Woodward, J. (2003). *Making things happen: A theory of causal explanation*. New York: Oxford University Press.

Wright, C., Colombo, M., & Beard, A. (2017). HIT and brain reward function: A case of mistaken identity (theory). *Studies in History and Philosophy of Science Part C: Studies in History and Philosophy of Biological and Biomedical Sciences, 64*: 28–40.

Wright, L. G., Onodera, T., Stein, M. M., et al. (2022). Deep physical neural networks trained with backpropagation. *Nature, 601*(7894): 549–55.

Zuboff, A. (1981). The story of a brain. In Dennett, D., & Hofstadter, D. (eds.), *The Mind's I*. New York: Basic Books, pp. 202–11.

Acknowledgements

This work was partially done on the territories of the Kickapoo, Kaskaskia, Myaamia, Ogaxpa, and Osage peoples. Thanks to Keith Frankish, Robert Laurent, Timothy Luft, and, especially, Marcin Miłkowski and Oron Shagrir for helpful comments on drafts of this Element. Thanks to Andy Clark, Nir Fresco, Corey Maley, Peggy Seriès, and Mark Sprevak for helpful conversations on this topic. This work was supported in part by a University of Missouri–St. Louis sabbatical leave to GP during Spring 2023. Any opinions, findings, conclusions, and recommendations expressed in this work are those of the authors and do not necessarily reflect the views of the University of Missouri–St. Louis.

Cambridge Elements ≡

Philosophy of Mind

Keith Frankish

The University of Sheffield

Keith Frankish is a philosopher specializing in philosophy of mind, philosophy of psychology, and philosophy of cognitive science. He is the author of *Mind and Supermind* (Cambridge University Press, 2004) and *Consciousness* (2005), and has also edited or coedited several collections of essays, including *The Cambridge Handbook of Cognitive Science* (Cambridge University Press, 2012), *The Cambridge Handbook of Artificial Intelligence* (Cambridge University Press, 2014) (both with William Ramsey), and *Illusionism as a Theory of Consciousness* (2017).

About the Series

This series provides concise, authoritative introductions to contemporary work in philosophy of mind, written by leading researchers and including both established and emerging topics. It provides an entry point to the primary literature and will be the standard resource for researchers, students, and anyone wanting a firm grounding in this fascinating field.

Cambridge Elements ☰

Philosophy of Mind